Typology of Industrialization Processes in the Nineteenth Century

FUNDAMENTALS OF PURE AND APPLIED ECONOMICS

EDITORS IN CHIEF

J. LESOURNE, Conservatoire National des Arts et Métiers, Paris, France
H. SONNENSCHEIN, University of Pennsylvania, Philadelphia, PA, USA

ADVISORY BOARD

K. ARROW, Stanford, CA, USA
W. BAUMOL, Princeton, NJ, USA
W. A. LEWIS, Princeton, NJ, USA
S. TSURU, Tokyo, Japan

Fundamentals of Pure and Applied Economics is an international series of titles divided by discipline into sections. A list of sections and their editors and of published titles may be found at the back of this volume.

Typology of Industrialization Processes in the Nineteenth Century

Sidney Pollard
Bielefeld University, Federal Republic of Germany

A volume in the Economic History section
edited by
P. David, *Stanford University, USA*
and
M. Lévy-Leboyer, *Université Paris X, France*

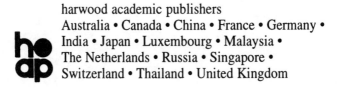

harwood academic publishers
Australia • Canada • China • France • Germany •
India • Japan • Luxembourg • Malaysia •
The Netherlands • Russia • Singapore •
Switzerland • Thailand • United Kingdom

First published 1990
Second printing 1996

Emmaplein 5
1075 AW Amsterdam
The Netherlands

Library of Congress Cataloging-in-Publication Data
Pollard, Sidney.
    Typology of industrialization processes in the Nineteenth century/
Sidney Pollard.
        p. cm. — (Fundamentals of pure and applied economics,
    ISSN 0191–1708; v. 39)
    Includes bibliographical references.
    ISBN 3-7186-5007-X
    1. Industrialization—History—19th century—Case studies.
    2. Economic development—History—19th century—Case studies.
    3. Industry and state—History—19th century—Case studies.
    I. Title.   II. Series.
    HD2321.P64   1990
    338.94′009′034—dc20

                                                                89-24670
                                                                CIP

Contents

v

Introduction to the Series

Drawing on a personal network, an economist can still relatively easily stay well informed in the narrow field in which he works, but to keep up with the development of economics as a whole is a much more formidable challenge. Economists are confronted with difficulties associated with the rapid development of their discipline. There is a risk of "balkanization" in economics, which may not be favorable to its development.

Fundamentals of Pure and Applied Economics has been created to meet this problem. The discipline of economics has been subdivided into sections (listed at the back of this volume). These sections comprise short books, each surveying the state of the art in a given area.

Each book starts with the basic elements and goes as far as the most advanced results. Each should be useful to professors needing material for lectures, to graduate students looking for a global view of a particular subject, to professional economists wishing to keep up with the development of their science, and to researchers seeking convenient information on questions that incidentally appear in their work.

Each book is thus a presentation of the state of the art in a particular field rather than a step-by-step analysis of the development of the literature. Each is a high-level presentation but accessible to anyone with a solid background in economics, whether engaged in business, government, international organizations, teaching, or research in related fields.

Three aspects of *Fundamentals of Pure and Applied Economics* should be emphasized:

—First, the project covers the whole field of economics, not only theoretical or mathematical economics.

—Second, the project is open-ended and the number of books is not predetermined. If new and interesting areas appear, they will generate additional books.

—Last, all the books making up each section will later be grouped to constitute one or several volumes of an Encyclopedia of Economics.

The editors of the sections are outstanding economists who have selected as authors for the series some of the finest specialists in the world.

J. Lesourne *H. Sonnenschein*

Typology of Industrialization Processes in the Nineteenth Century

SIDNEY POLLARD

Bielefeld University, Federal Republic of Germany

INTRODUCTION

Among economic historians, the "process of industrialization" is a technical term. It is not to be understood as implying that there was no industry before its onset. On the contrary, in many European countries as well as in some countries outside Europe industrial activity had existed for many hundreds of years before "industrialization" set in. Nor is it to be understood to mean that the development of industry was necessarily central to it. Very often it was, but in some countries, it was the transformation of the primary sector, agriculture or forestry in particular, that was the key feature. In more recent years, the technical transformation or modernization of the services sector has sometimes played an important part.

Essentially, what is implied by the term "industrialization" is a major economic change, affecting ultimately all sectors of the economy, in which output is increased, efficiency raised and new products are being produced, by methods involving new technologies and the enlarged use of capital. These, in turn, also necessarily involve improved means of transport, of communication and of trading, the application of science, the transformation of relations between employer and employed, the building of factories and of other large units of production such as mines or shipyards, a minimum standard of literacy, the adaptation of the legal system to strengthen the security of private property, or specifying the limits of public property, and other related features.

Most of these are rather vague specifications lacking precise

1

indications of numbers on percentages. Thus improved technologies, some capital, better roads or a higher level of literacy were achieved also in other ages. In the end, the concept of "industrialization" is tied to a specific and unique phase in mankind's progress recognisable by certain structural characteristics (Kuznets 1966) and a certain level of national income (Bairoch 1981). This phase began with the "industrial revolution" in the second half of the eighteenth century in Great Britain.

The British industrial revolution was the first example of an industrialization process in the accepted sense. Sometimes described as the "classical" industrial revolution (Coleman 1962) this term itself has been subject to criticism (Fores 1981). In particular, by analogy with political revolutions, it has been said that the term conveys the misleading impression of a "sudden, discontinuous change".[1]* However, in other fields, as for example astronomy, no indication of any particular time span is intended by "revolution" (Porter and Teich 1986), and it need not be assumed here. In any case, the term has now become so widely used (Toynbee 1884, Ashton 1948, Clark 1953, Mantoux 1961, Deane 1965) that it may be accepted as a technical term without reference to its two component parts considered singly. Indeed, some attempts have been made to apply the term to the sixteenth century, the thirteenth century, even to prehistory to designate major technical breakthroughs (Coleman 1962), but these have been widely rejected because they do not refer to the singular phase of industrialization.

Beside Great Britain, a limited number of other countries, notably Belgium and Germany, are sometimes said to have experienced an "industrial revolution". The implication that the British experience was repeated there with sufficient similarity to warrant the same description has not been without its critics. It is part of a much wider debate how far the process of industrialization in different countries may be taken to be a similar or like path, or alternatively whether there existed quite different and distinct models or types of industrialization.

There is no dispute about the similarity of the starting position, and of the final arrival zone. Economies before industrialization were characterised by low output per head, a very large agricultural sector, and strict limits set to economic growth per head because at the given

*See Notes section at the back of the book.

level of technology, natural resources showed diminishing returns (Rostow 1960, chapter 2, 3, Wrigley 1988). Despite enormous differences in detail, ranging, for example, from the capitalist agriculture of Britain to the serfdom of Russia, there were, at least within Europe, sufficient basic similarities to qualify for a common "style" within the framework of Spiethoff's (1953) Gestalt theory. The similarities were certainly greater after industrialization, with strong signs of convergence also among industrialized countries outside Europe, such as the USA, Australia or Japan, and some convergence even among those still in the early stages of the process (Moore 1979).

The path between the "traditional" and the industrialized forms of society, however, which is of interest to economic historians and to development economists alike, shows considerable divergences between different countries. Are these modifications of the same phenomenon, closely related phenomena, or different models altogether? The problem is faced by many sciences, but most have developed clear systems of classification. There is, for example, no dispute among zoologists as to the relationship between, say, a St. Bernhard dog and a dachshund, or between either and a fox or a cat. In economic history such agreed classification criteria do not exist. While the debate has therefore something of a semantic character, it is not without a real underlying issue.

Three broad types of views may be discerned. The first sees the process of industrialization as repetitive in each country. The second sees it as a form of modified repetition. The third considers that the paths in different countries followed lines so different as to constitute a set of quite distinct types or models of the process of industrialization.

Karl Marx may be reckoned among the supporters of the first view. "The country that is more developed industrially", he wrote in *Capital*, "only shows, to the less developed, the image of its own future".[2] Among modern adherents of this view may be cited Landes (1969), Hughes (1970 chapter 5), and the authors of stage theories such as Hoffmann (1958) and Rostow (1960, 1978). The latter, indeed, claims to be able to identify the timing of analogous phases or stages of the same industrialization process in each country. One implication might be that it would be sensible to draw up national income statistics on the same basis for each country (Fremdling and O'Brien 1983, O'Brien 1986, Bairoch 1976).

A theory of modified repetitiveness has been put forward by

Gerschenkron (1966b). According to this, the delay in the industrialization process found in the later comers, their "backwardness" compared with Great Britain, which is due to their lack of some significant factors, has to be made up by some alternative or substitute processes. These may include a very fast spurt, the formation of very large industrial units from the start, the use of bank capital for fixed investments and, for later industrializers still, the use of capital accumulated by the State, and a much larger role for the state altogether. Apart from these enforced differences, the process is, however, essentially similar and imitative, at least among the European industrialisers. A modified version of this schema has been proposed by Pollard (1981), emphasising the passage of time on the necessary technology of the latecomers at similar stages, and the effects of the simultaneous introduction of new technologies such as the railways on countries at very different stages of their development. In this way, sequence and interdependent effects were modified, while still leaving recognisably the same process in each country.

Another view which may be classified as modified repetitiveness is that which stresses the differences between the early industrialisers as a group, and the European or extra-European "periphery" as a group. The latter may be held back altogether for long periods (Wallerstein 1974, Berend and Ranki 1982), though a transfer from one category to the other may in some cases be possible.

The implication of several of these views is that the industrialization in each country is affected strongly by what goes on in other, more developed countries — in other words, that in some sense, the industrialization at least in Europe and in North America must be seen as a single process, which will be obscured if countries are studied individually on their own. Such concepts have been put forward by Sombart (1927–8), Pollard (1973) and Jones (1981, 1988).

A third group of historians denies the usefulness of a single model, however modified, and prefers to consider industrialization to have taken place according to several different models or paths (Ashworth 1977, Cameron 1985). One method to arrive at such a result is derived from anthropology and sociology. It consists of listing the largest possible number of attributes[3] and observing whether, by the statistical method known as disjoint principal components calculation, certain bunchings of standard patterns would emerge showing close similarities within each pattern, but being clearly marked off from the others (Morris and Adelman 1988, also Chenery 1960, Chenery and Syrquin).

In the event, five different growth paths were discerned by Morris and Adelman: early industrializers, enjoying the advantages of an efficient agricultural sector: a second wave of large European industrializers, using State power and protection; primary producers with abundant land; primary producers under land-scarce conditions; and the balanced-growth path taken by several smaller, but rich European countries.

Concentrating on the differences between countries permits the use of the comparative method, as favoured by Marc Bloch (1953, also Wewell 1967, O'Brien 1986). One such comparison which appears frequently in the literature is that of balanced versus unbalanced growth. As far as the industrial sector itself is concerned, there is a widespread consensus that during industrialization, individual industries typically were modernised and expanded at very different rates. One or more "leading sectors" would pull up the others in what was clearly a form of unbalanced growth (Rostow 1963b, Hirschman 1985). There is, however, much less agreement on the path normally taken by agriculture.

Observation of the industrialization process in Europe, North America and Japan has convinced many that a highly developed, modernized agriculture is a necessary precondition for the onset of industrialization (Milward and Saul 1973, 1977, Jones and Woolf 1969, Eichner and Witt 1964, Gould 1972, chapter 2, Hughes 1970, but see Treblicock 1981). This would imply a balanced growth pattern in which rising agricultural productivity provides both rural market and a labour supply for the expanding industrial sector. An alternative, strongly unbalanced growth amounting to a dual economy, is favoured by Lewis and by Ranis and Fei (Lewis 1954, 1958, Ranis and Fei 1961, Fei and Ranis 1969). In this concept it is a backward, underemployed agrarian and traditional economy which furnishes labour to the modern sector. Such limited econometric studies that have been made do not come down clearly on either one or the other side of the argument (Geary 1988).

Here it is not necessary to decide whether the differences between the major countries engaged in the process of industrialization are best characterised as following paths that were modifications of one type, or were sufficiently different from each other to warrant classification as different species. The presentation will be descriptive, with emphasis on the main characteristics of each path followed.

Regional Patterns of Industrialization

Before proceeding to present a typology on a national basis, it is necessary to note that this approach may hide some of the key dynamics of the process. Industrialization in fact took place in every country on a regional basis, and a regional typology, could it be developed, might well yield more information than can be derived from the larger national units.

That industrialization was a regional process framework was clear to all observers, contemporary and later. Nor is this merely to be understood as an irregular or stochastic geographical bunching. The regional structure provided many advantages for the new industrialism, quite apart from being a natural result of such exogenous factors as the location of coal or of metallic minerals. Industrialization in regional concentrations reduced transport costs, improved the markets for goods, capital and labour, made possible cost-saving institutions and, especially in countries with scarce resources, concentrated the available capital and entrepreneurship into a minimum critical mass. Technical and other traditions and even legal systems, such as those relating to landownership, were frequently regional. Industrialised regions played, economically speaking, a role similar to and had far more in common with similar regions in other countries than with neighbouring non-industrialised regions (Pollard 1980, 1981, Lebrun 1979, Pounds 1985, Perloff, Dunn, Lampard and Muth 1980, North 1955, Fremdling and Tilly 1979, Söderberg 1985, Williamson 1965).

Nevertheless, history continues to be written on a national basis. Among the main reasons for this is that most statistics are available for national or country units only, and that economic policies on such matters as tariffs, taxation, monetary developments or monopolies have always been conducted on a national basis (Kuznets 1951, Trebilcock 1981). Even where, for historical reasons, as in Germany or Italy, smaller units had an independent political existence with their own statistical data (Tipton 1976, Borchardt 1968, Zamagni 1978) it is frequently difficult to put these on a common base.

However, the regional factor must not be altogether neglected. It imposes a certain caution in interpreting national "types". Industrialization, for example, may proceed very similarly in two or more regions; but because one is almost coterminous with the national boundaries, e.g. in Belgium, while another is swamped in national statistics by large undeveloped regions, as in Austria–Hungary or Germany, misleading conclusions may be drawn from purely national information about the character of a country's industrialization process.

1. GREAT BRITAIN AND WESTERN EUROPE

1.1. Great Britain: The Pioneer Industrial Revolution

Slow Rate of Growth
Possibly the most marked characteristic of the British process of industrialization, as compared with others, is its slow rate of growth. The data base, particularly for the eighteenth century, is fraught with uncertainties, and even its key element, the size and growth rate of the population, is not without dispute. Nevertheless, most historians now agree that the first decades of the industrial revolution, dated conventionally to start somewhere between 1760 and 1780, saw at most a scarcely perceptible rise from the previous very modest rate of economic growth, and even most of that was swallowed up by the growing population, so that the increase per head was practically negligible. Only in the last years of the century did a very slight acceleration begin which was then continued more markedly to reach quite respectable growth figures by the middle of the nineteenth century.

In comparison with the overall growth rates achieved by the British economy, as well as by other countries in their early industrialization phases in the nineteenth century, let alone in comparison with the growth rates of the later twentieth century, "industrial revolution", or transformation, does indeed seem, at first sight, to be a misnomer. Those who lived through that phase of history can hardly have counted the growth of their incomes as one of its more noticeable features.

Two sets of explanation for the British peculiarity of very slow growth in the period of industrialization may be offered. The first is to note that Britain at the time was at the technical frontier. Unlike the

TABLE I

British Economic Growth Rates 1700–1841, in per cent per annum[a]

| | Harley | | | | Crafts | | |
	Income	Income per capita	Total productivity		Real output	Real output per head	Real commodity output
1700–1770	0.56	0.27	0.00	1700–1760	0.69	0.31	0.64
1770–1815	1.31	0.33	0.25	1760–1780	0.70	0.01	0.61
1815–1841	2.23	0.86	0.71	1780–1801	1.32	0.35	1.35
				1801–1831	1.97	0.52	2.18

[a] Sources: Harley (1982), p. 287, Crafts (1987), p. 246.

later followers, the lead economy has no free space for catching up, and can advance only as fast as technological progress itself (Maddison 1982). At the same time, again unlike the later industrializers, British industries did not have to fear the competition of more advanced established industries abroad. No matter how relatively inefficient and sluggish their progress, they were still at every point ahead of the foreign competition.

Moreover, technical progress itself in the later eighteenth century was exceedingly slow compared with that of more recent years, though it might have appeared very rapid, almost miraculous, compared with former ages. The link with science — itself advancing only slowly — was tenuous at best (Musson and Robinson 1969). Inventors worked largely by hunch, R & D expenditure was negligible, even the expertise of key craftsmen, such as machine builders, still had to be created. Few men, in absolute numbers, concerned themselves with innovation, and, as always, progress in one line had to wait for innovation in another: Watt's steam engine could be built only after John Wilkinson had perfected a method for the true boring of cannon, and cotton spinning could take off on its breathtaking expansion path only after the cotton gin for separating out the seeds from the fibres had been perfected.

While later comers could take over their technology ready made, the pioneer inevitably took wrong turns and suffered from numerous false starts. Many canals proved unnecessary in view of the impending railway network, while railway building was needlessly costly because of the failure to see early on that a national network would have to emerge. There was, in any case, no national steering or planning. Initiatives were private — the State, in its ignorance of what was emerging, being at times inadvertently helpful, at others the reverse. It is uncertain whether this absence of State action on balance retarded or accelerated developments.

A second consideration is possibly of even more profound importance. The early stages of industrialization were marked by the rapid transformation of several small corners of the economy, but these were at first too limited to affect output or income figures on a national basis. Only after even wider circles of the economy had begun to be transformed would this show up substantially on the overall growth rate. Accelerated growth in the nineteenth century was thus not due to faster growth in any given sector, but to a growing number of sectors having joined the process of mechanisation and technical progress, as well as to the necessarily increasing weight of the fast growth sectors in the economy.

Total Factor Productivity, shown in Table IIA, is a conventional measure showing the increase in output after the contributions of additional capital and labour inputs have been deducted. It thus approximates to that part of economic growth due to better technology and organisation. It will be seen that while the annual TFP growth for textiles, iron and transport was quite high for the period 1780–1860, it was swamped nationally by the low contributions of other sectors. Table IIB shows the effect of the increasing weight of the growth sectors within British industry in this period.

Agriculture, however, remained the largest single occupation, employing 35.9% of the occupied population in 1801 and 21.7% still in 1851 (Deane and Cole, 1967, p. 142, but see Wrigley 1986). The contribution by agriculture to the industrialization process is much in dispute. According to Crafts (1987) its TFP was rising in 1761–1800, at 0.2% a year, at the same rate as that of industry, while in the period 1801–61, rising at 0.9–1.0% a year, it was far ahead of the industrial growth rate (also O'Brien 1977). Similarly, O'Brien and Keyder (1978) attribute Britain's good showing as against France in the nineteenth century entirely to her superior productivity in agriculture.[4] But this, it has been argued, proves too much. It would mean that Britain would have done better to specialise in agriculture rather than become the workshop of the world (Williamson 1987). While Williamson's growth rate for agriculture at a little above zero may be too low, a consensus of around 0.4% a year to the mid-nineteenth century seems to be emerging (Mokyr 1987, Floud and McCloskey 1981, chapter 4 and 10, and pp. 114–6).

Agriculture did remain a population reservoir for the industrial towns. This did not necessarily mean that the agricultural population declined, but that the surplus at least was being syphoned off by the industrializing sectors (Wrigley 1986, Williamson 1985, Pollard 1978). The push-effect away from the countryside, particularly after enclosures, may have been reinforced by the pull-effect of higher wages in towns (Williamson 1985, Crafts 1985b, Chambers 1953). Certainly, agricultural wages rose faster, and to higher levels, in regions near the industrial towns than in purely agrarian regions (Williamson 1987).

Given the uncertainties about the relative productivity in agriculture and industry, it must also remain in doubt whether the reallocation of resources from agriculture to industry which took place, contributed much to the British growth rate in those years. Similarly, in view of the advanced traditional industrial structure of the country before the industrial revolution, it is not clear whether transfers within the industrial sector will have raised total productivity by an appreciable amount, other than by the gains through improved technology. Both these factors, however, were likely to have contributed to the productivity spurt of later industrializing countries which began their process with a much larger agrarian sector, and a much less advanced industry, than did Britain.

Capital Investment and War

In Britain, as elsewhere, industrialization was marked by an acceleration of capital formation. This was in line with expectations of theory, while empirical studies made since the Second World War showed that countries that were fully industrialised had investment ratios of at least 12% of national income, whereas countries with traditional economies generally kept to a ratio of below 5% of their much lower national income. What sometimes became known as the Lewis–Rostow thesis therefore maintained that industrialization implied an increase in the investment ratio from around 5% to around 12% (Rostow 1963, Introduction).[5]

In the British industrial revolution with its much gentler and lower rate of acceleration, no such drastic increase need necessarily have been expected. It would, moreover, have to be born in mind that the British economy started from a higher level, being the source of finance for much of the world trade of the day as well as having a more sophisticated economic structure than current primary producers.

In the event, its has proved extraordinarily difficult to establish the actual rates of investment during the British industrialization, either in absolute terms or as a proportion of national income. At one extreme Deane and Cole estimated that the investment rate, standing at 5–6% of the national income in the 1790's, might have reached 7% by the turn of the century but would certainly not have increased in the war years thereafter. Only the railway boom of the 1830's brought a further noticeable increase, by perhaps 2% (1967, p. 263, also Deane 1972). Against this, Feinstein found Gross Fixed Capital Formation as a proportion of Gross Domestic Product (GDP) to have risen from 7% in the 1770's and 1780's to 10–11% thereafter, total domestic

TABLE II

Sectoral Economic Growth in Britain, 1770–1860

(A) Annual Productivity Changes by Sector 1780–1860[a]

	Rate of growth of TFP[b] (% per year)	(2) Weight	Contribution to overall TFP growth (% per year)
Cotton	2.6	0.07	0.18
Worsted	1.8	0.035	0.06
Woollens	0.9	0.035	0.03
Iron	0.9	0.02	0.018
Canals and railways	1.3	0.07	0.09
Shipping	2.3	0.06	0.14
Weighted sum of modernized	1.8	0.29	0.52
Agriculture	0.45	0.27	0.12
All other sectors	0.65	0.85	0.55
Total		1.41	1.19

(B) Value Added in British Industry 1770–1831 (£m, current)[c]

	1770	%	1801	%	1831	%
Cotton	0.6	(2.6)	9.2	(17.0)	25.3	(22.4)
Wool	7.0	(30.6)	10.1	(18.7)	15.9	(14.1)
Linen	1.9	(8.3)	2.6	(4.8)	5.0	(4.4)
Silk	1.0	(4.4)	2.0	(3.7)	5.8	(5.1)
Building	2.4	(10.5)	9.3	(17.2)	26.5	(23.5)
Iron	1.5	(6.6)	4.0	(7.4)	7.6	(6.7)
Copper	0.2	(0.9)	0.9	(1.7)	0.8	(0.7)
Beer	1.3	(5.7)	2.5	(4.6)	5.2	(4.6)
Leather	5.1	(22.3)	8.4	(15.5)	9.8	(8.7)
Soap	0.3	(1.3)	0.8	(1.5)	1.2	(1.1)
Candles	0.5	(2.2)	1.0	(1.8)	1.2	(1.1)
Coal	0.9	(4.4)	2.7	(5.0)	7.9	(7.0)
Paper	0.1	(0.4)	0.6	(1.1)	0.8	(0.7)
	22.9		54.1		113.0	

[a] Source: Floud and McCloskey (1981) p. 114, modified by Crafts (1987) p. 249.
[b] TFP = Total Factor Productivity.
[c] Source: Crafts (1985) p. 22.

investment from 8–9% to 12–13%, with a relapse in the war years, and total investment (including investment overseas) from 8–10% to 13–14% (1972, also Feinstein and Pollard, 1988, Part II). These rates would correspond fairly closely to the Lewis–Rostow assumptions.

The discrepancies are explained in part by the defective sources of information, particularly as regards the two very important sectors of agriculture and housing. In some other cases, they arise out of problems of definition as to what constitutes investment (Feinstein and Pollard 1988).

Differences in interpretation also arise because of the varied forms in which capital appeared, and the multitude of its sources. Unlike many later industrialisers, Britain received very little capital from abroad; such borrowings as had existed from Holland in the eighteenth century had largely been repaid, so that the sources of capital had to be internal. There is some debate as to whether capital was adequate or in short supply.

Broadly speaking, capital was required in three forms. For large items of the infrastructure, such as roads and canals, later also railways, as well as for the ships and "factories" of the trading companies, joint-stock organisation could tap the savings or large numbers of wealthy people from different walks of life. For circulating capital, needed by industrial firms as well as by trading firms at home and abroad, the banks could provide credit, in part by channelling it from the agrarian surplus areas to the industrializing deficit regions. As for fixed capital in agriculture and industry, this had largely to be provided by owners and partners themselves, but might be drawn also from landowners via mortgages or the provision of permanent structures in farms, mines or even factory buildings. However, fixed capital was of less importance in the early stages of industrialization (Crouzet 1972b, Pollard 1972). There might thus be acute shortages of capital for specific purposes or in particular times, and there might be pressure on resources, but there is no sign that British industrialization was held up in a serious way by an overall shortage of capital. Interest rates were low by the standards of the eighteenth century, as for that matter by current standards, and they had a falling tendency.

There is one exception to this generalisation. It has been alleged by Williamson (1984, 1987) that the requirements of the wars against France, lasting with brief interruptions from 1793–1815, placed such a heavy burden on the economy that, by absorbing a large share of savings, they "crowded out" private investment to the point that economic growth was seriously held back. It is this, according to him which prevented the rise in the British standard of living until about 1820, after which vigorous growth of output and incomes could at last taken place.

There is some evidence that even in eighteenth-century wars, Governments diverted much of the available savings to war purposes with noticeable effects on diminished productive investments in the private sector (Ashton 1959). As an explanation for the slow rise in output and living standards in the French war years, however, the thesis has not found widespread support. It is said, in refutation, that the quantities of new borrowings by the Government in real terms have been greatly exaggerated, that possible increases in the savings ratio have been ignored, and that the men in the forces consisted in part of Germans, Dutchmen, Scandinavians as well as Irishmen, reducing to that extent the loss to the British labour force. The case for "crowding out" therefore remains uproven (Mokyr 1987, Crafts 1987, Heim and Mirowski 1987).

Structural Transformation
The most obvious outward signs of the British industrial revolution, and the development most stressed by contemporaries and foreign visitors, was the transformation of certain key industries. In these, technical changes allowed output to be increased and costs to be cut, which further increased markets and led to further technical improvements in a continuing virtuous spiral. The industries affected first were a mixed bunch, and it is not easy to classify them.

Prominent among them in the consumer sector were the textile industries. This is not entirely surprising, since textiles formed much the most important manufacturing sector before "industrialization" while, conversely, clothing was the most important item of expenditure in the household of the poor after the elementary needs of food and shelter had been met. There was a certain bunched sequence: after the early mechanisation of silk throwing, a small industry which had no linkages elsewhere, it was cotton that was mechanized first, followed by woollens and worsteds, and finally by linens. Mechanisation started in the early stages of production with carding and spinning; mechanical weaving followed some 40–50 years later, while the final stage, tailoring, was not mechanised until the second half of the nineteenth century, as a result of the development of the (American) sewing machine.

The inventors of cotton spinning machines, Arkwrigth, Hargreave and Crompton, making their contributions in the 1760's and 1770's, were only the first in a widening stream of inventors and improvers, covering ever extending areas. The rate of expansion of these sectors was astonishing. Thus retained raw cotton imports, a convenient measure of cotton textile production, rose (in million lbs a year) from 4.2 in 1772–4 to 99.7 in 1815–17 and to 621 in 1849–51, or about 150-fold. Value added rose much less because of the cost reductions, averaging (£ million) 0.6, 21.7 and 30.2 in the same years, which still amounted to a fifty-fold increase in monetary terms. In woollens and worsteds the quantity of wool consumed rose from 105 million lbs a year in c. 1805 to 241 million lbs in 1850–4, linen raw material imports rose from an index of 74 (1770) to 191 (1845) and silk from an index of 87 (1770) to 521 (1847) (Deane and Cole 1967). At the same time, applied chemistry raised dyeing and bleaching far beyond their former capacities.

No comparable breakthroughs occurred in other consumer goods, unless we take entirely new products such as gas for domestic lighting, or the new means of transport. At the same time, few sectors remained entirely without progress. Thus in pottery, Josiah Wedgwood improved the quality of his wares by chemical research and by a revolution in internal organisation; porter brewing became a mass-production industry; immense improvements in detail, not least in the motive force in the workshops, were made in the production of metal goods, while printing and paper-making experienced revolutions of their own.

Among capital goods, it was the production of iron which showed the most remarkable increase. After the discovery of coke smelting by Abraham Darby in 1709, which became widely applied from the 1750's onward, and the puddling process, invented in 1783 by Henry Cort, the mass production of relatively cheap iron could begin. Output of pig iron rose from 44,000 tons in 1775 to 1,396,000 tons in 1841, or more than thirty-fold, while continuing technical progress ensured that annual improvement in productivity was 0.9%, compared with 2.6% in cotton, 1.8% in worsteds and 0.9% in woollens.[6]

The provision of industrial motive power was a key element in the expansion of the capital goods sector. The steam engine, for which the significant technical breakthrough was James Watt's separate condenser of 1769 to improve the traditional Newcomen engine, became symbolic for the new industrialism. But it should not be forgotten that other sources of power, notably the water wheel, continued to be improved and extended also (Musson 1978). The steam engine, however, was important also in helping forward engineering, which was revolutionised by several men of genius, such as Bramah, Maudslay, Clement and Whitworth in this period.

Lastly, coal mining should be mentioned among the fast growers. Its output index rose from 15 in 1770 to 46 in 1815 and 100 in 1841, compared with 7 and 29 and 100 for metals for the same years (Harley 1982). Coal mining, also, was a seedbed of engineering progress: both the steam engine in its earlier form and the railway were developed within that sector. In Britain, coal was both a consumer good and a source of energy used in industry: the geography of British industrialization is coterminous with the geography of British coal.

Technical or organisational needs in most of these innovating industries led to the establishment of units of large size. In coal mining, the largest mines in the North-East as well as in some other regions came to employ several hundred men each together with capital in the form of pumping and winding engines, rail systems, lifting gear and buildings on the surface which quite regularly exceeded £20,000 per pit in the early 19th century. Cotton and other textile mills generally required less capital, but also employed many hundreds of workers each, mainly women and children. Large complex ironworks came to reckon their total employment in thousands, while canals constituted large concentrations of capital and acclimatised new classes of owners to the vagaries of share owning and of the stock exchange.

The "factory" became symbolic for the new industrialism. Its technical justification was frequently the need for an efficient central motive power, or for an orderly progress of the material from process to process through the works. The factory also permitted closer surpervision of the workforce, including quality control, the saving of fuel and raw material and the prevention of embezzlement. Correspondingly, the factory was felt by many of its workers to impose a severe limitation of their freedom, to be an impersonal form of work organisation as well as being an unhealthy and frequently dangerous place of work. Whether it led to harder work or to longer hours, whether it placed more work on the shoulders of women and small children than had been customary in Britain before, is in dispute. It certainly made the working hours more regular and more tedious.

Yet the spread of the "factory", even if we include all large units, including mines, ironworks and shipyards in that description, was far from universal. As late as 1850, cotton spinning was the only sector even among the new industries in which a majority of workers were found in large units. Everywhere else the small workshop, frequently still containing craftsmen of the traditional type, predominated. Even by 1870, official returns indicate that outside cotton and woollen mills with their associated dyeing and bleach works, the iron industry and engineering where there were 645,000 HP installed altogether, steam power was still extremely limited in Britain: there were only 80,000 HP in no fewer than 51,548 establishments in the rest of industry, or a little over 1½ HP per establishment.[7] Many more people were still employed in "hand" technology than in machinery technology even in mid-Victorian Britain, a generation after the industrial revolution was alleged to have run its course (Samuel 1977).

Nevertheless, the modernized sectors were the dynamic elements in the economy, the sectors which distinguished the British economy and gave it its unique place in the world, and the sectors which other countries, in their different ways, attempted to copy.

Causes

The British industrial revolution as the first process of industrialization, without any direct antecedents, may therefore be said to have been the origin of a unique worldwide and irreversible phenomenon. The question of its causes has therefore been much debated. Traditionally, the emphasis has been on the supply side, but the demand side as cause has also found its defenders.

The supply side approach tends to emphasise one or more independent or exogenous factors which, after a period of growth, came together to transform changes in quantity into changes in quality, or, in more modern terminology, to reach a critical mass which set in motion a fundamental transformation in the economy. Among the more widely enumerated factors are the following:

(a) The accumulation of capital. This is associated with a general rise in incomes and wealth, but also with distributional change and institutional innovation. More wealth flowed into the hands of the "investing" middle classes rather than the "spending" landed or poorer classes. Banks, and the ploughing back in rising capitalist enterprises themselves, provided mechanisms for turning savings into capital.

(b) Innovations: these include the technical inventions, larger and better organisation, also perhaps regional concentration.

(c) Resources: coal, iron and other minerals were found, often just out of reach of current technology thus encouraging technical innovation. Locational factors, such as good coastlines, a position astride the increasingly important Atlantic routes, and a good supply of skilled labour and able entrepreneurs are also included here (Fohlen 1971).

(d) Political and intellectual elements. These include: internal peace, political freedom and a class constellation favouring "laissez-faire" policies at home while keeping on protection abroad until Britain did not need it any longer. The command of the seas and the acquisition of a colonial empire are sometimes enumerated. The rise of science, of secularism and individualism, appropriate changes in the law, and free social mobility have also been included.

(e) Among a group of further miscellaneous causes were the freedom from invasions while the continent was racked by war, the decline of the plague, and the "English genius" for compromise, for practical tinkering and for bartering (Hartwell 1965).

Defenders of supply-based theories find much fault in detail with the demand theory of causation (Mokyr 1985b). There is the further question why the changes occurred solely in Britain rather than in France, which provided a much larger market and would have had access to the same technology (Crafts 1985a). Rising demand, moreover, especially if caused by rising population, had in previous ages not been considered a favourable factor, but one threatening survival (Wrigley 1988); if it turned out to be a favourable factor this time, it must be due to altered supply conditions.

The emphasis on demand goes back to a classic article by Elizabeth Gilboy (1932) but it still has its vigorous defenders today (Floud and McCloskey 1981, chapter 3). According to them, the rise in home demand, inasmuch as it is distinct from the increasing incomes derived from the industrialization process itself, can be traced back to two major sources: the successful expansion of agricultural output in the first half of the eighteenth century, that is, immediately antecedent to the "take-off", and the internal redistribution of income.

Rising agricultural output, ahead of population increase to the mid-century, increased the total national product in real terms and, by lowering food prices, redistributed incomes in favour of the non-agrarian population. There was also a general rise in the share of income going to the middle and middling classes: these provided a mass market for manufactured goods in place of the former very limited absorptive capacity of the poor on the one hand, and the luxury market which did not lend itself for mass production, on the other. Besides, it can be shown in detail that most of the key firms and industries found the great bulk of their additional markets at home rather than abroad (Eversley 1967).

Foreign demand as trigger has the advantage that it is a truly exogenous factor and thus does not suggest the feat of pulling oneself up by one's own bootstraps. In part, it derived from British entrepot trade and the growth of processing industries, like tobacco curing and sugar refining, on the way between colonies and European markets. In part, exports of goods like cottons and woollens to Europe merely displaced German linens or French wollens, and thus need no *deus ex machina* explanation of a rising demand.

Exports, particularly of the key products of cottons, woollens, iron and iron goods,[8] expanded in irregular waves. In periods of rapid expansion, such as the mid-1770's (35% increase for the quinquen-nium), the mid-1790's (45%) and the years around 1805–10 (23%) (Crouzet 1985, chapter 7, Floud and McCloskey, Chapter 5) it would be foolish to deny the influence of foreign markets on British industrial progress. This does not mean, however, that exports were significant in triggering the British industrial revolution (Findlay 1982).

The search for a monocausal explanation goes on. More widespread is the belief that a conjunction of both demand and supply factors was the most likely "cause" of the British industrial revolution. At the same time it should be emphasised that most current theories agree that British industrialization came about through positive rather than negative causation, i.e. in response to opportunities rather than in reaction to a crisis.

The International Setting
As the pioneer, the British economy was for a time the only industrialized one in existence; even later, when others had also begun to industrialize, it was still ahead of them. Its relationship with the rest of the world was therefore unique. Crafts (1985) attempted to establish the nature of this uniqueness by comparing some key data for Britain

with the "European norm", i.e. the average of other European countries, not at the same time, but *at the time when they reached British per capita incomes* in the given years. Table III, reproducing some of his findings, shows that Britain reduced her agricultural sector much earlier, inserting herself into an international division of labour more strongly to draw more of her food supplies from abroad, than other countries. While her employment in the industrial sector was higher, her income from industry was not, pointing to her larger services sector. Lastly, her much lower rate of investment should be noted, a consequence both of her slower growth rate and her investment abroad, which may be considered in part a substitution for investment in home agriculture or forestry.

Britain's unique intermediary trading position between the overseas world and Europe included the slave trade as well as the trade with her colonies. At one time it was thought that those provided much of the accumulation of capital necessary for industrialization, or, put briefly, that industrial capitalism was erected on the exploitation of black slaves.

Such views are no longer tenable today. Apart from monopolised trades, like that of the East India Company, trading, it can be shown, brought no more than normal profits. This is also true of the slave trade. The North American colonies, as Adam Smith had noted correctly, cost more in defence and administration than they ever brought in surplus (Floud and McCloskey 1981, Chapter 5, Engerman 1972). Even in the trade with the sugar islands, heavy investments and the naval expenditure have to be set against large profits so as to leave no positive balance: in one view, "if the West Indian colonies had been given away, British national income would have increased" (Coelho 1973, p. 254). The opposite view, on the basis of a Marxist analysis, holds that Britain gained from slavery and from her colonies by enjoying a wider division of labour (Inikori 1987). Both may well be true.

Social Consequences
Industrialization made some people rich. It also bore hard on others, particularly the workers in the new factory towns. Contemporary observers frequently expressed in official enquiries and in the literature, their horror at the conditions of the new industrial conglomerations, and a succession of protest movements by the people affected

TABLE III
Britain's Development Transition and the European Norm[a]

	1760		1800		1840	
	Britain	Europe	Britain	Europe	Britain	Europe
Urbanization	na	na	33.9	23.2	48.3	31.4
% Male labour force in agriculture	52.8	66.2	40.8	64.0	28.6	54.9
" " in industry	23.8	16.9	29.5	18.6	47.3	25.3
% Income in primary sector	37.5	46.6	36.1	44.8	24.9	37.2
" industry	20.0	21.3	19.8	22.0	31.5	25.2
Investment as % National Expenditure	6.0	12.2	7.9	12.6	10.5	14.4
Foreign capital inflow as % National Expenditure	na	na	0.6	0.5	-1.2	0.1

[a] Source: Crafts (1985) pp. 62–3.

bear witness to the same feeling. Overcrowding, unhygienic conditions, economic uncertainties, exploitation by factory owners and shopkeepers, long hours of work were among the most common complaints. These things are difficult to quantify, and were denied by other contemporaries. Nor is it easy to compare pre-industrial conditions on the land, or in the few large towns like London, Edinburgh or Bristol, with the later Manchester or Birmingham.

One area in which more precise statements are possible is the course of real wages. It lies at the heart of the "standard of living controversy", on which an immense stream of literature has accumulated which shows no sign of abating. In terms of real incomes in the period c. 1790–c. 1840 (the beginning and end-dates are themselves subject to debate) the "optimists", believing in improvement, have rather the better of the argument, but the "pessimists" can point plausibly to the deteriorating living and working conditions, at least as perceived by those who lived through them. No general agreement seems in sight, though there is a certain consensus that after 1770 and up to about 1820, real wages stagnated and may even have declined, while thereafter a rise was much more evident. After the mid-forties all observers are agreed on a marked improvement in the standard of living of the mass of the population (Mokyr 1988; Crafts 1985 chapter 5; Lindert and Williamson 1985). Only after the industrial revolution was over, did the real benefits to the people at large become clearly observable.

1.2. Belgium: The First Industrializer on the Continent

Industrial Change
Western Europe (together with the USA) was closest to Britain in economic and social structure. Among the " 'followers' or 'latecomers' who tried, during this period (1815–1850, S.P.), to emulate Britain and to introduce the new technology which she had invented"[9], Belgium was the first. The leading industries, coal mining, metallurgy, engineering and the textiles, were much the same, though the sequence and relative sizes showed some distinct features. The new techniques adopted were those pioneered in Britain. As Table IV shows, Belgium was not far behind Britain in overall development, and its industrial rate of growth, though slower than the British, was still remarkable.

Belgium was well endowed with coal. Coal output expanded in the

24   SIDNEY POLLARD

TABLE IV
Levels of Industrialization in Some European Countries, 1750–1860[a]

| | Per capita, Index UK 1913 = 100 | | | Total industrial potential[b] UK 1913 = 100 | | | |
	1750	1800	1860	1750	1800	1830	1860
United Kingdom	10	16	64	2.4	6.2	17.5	45.0
Belgium	9	10	28	0.4	0.7	1.3	3.1
Switzerland	7	10	26	0.2	0.4	0.8	1.6
France	9	9	20	5.0	6.2	9.5	17.9
Germany	8	8	15	3.7	5.2	6.5	11.1

[a] Source: Bairoch (1982) pp. 281, 292.
The levels are intended to indicate industrial production.
[b] In each case the average of three years, of which the year given is the middle one.

eighteenth century with the aid of Newcomen engines which were installed almost as early as in Britain: the first Boulton and Watt steam engine dates from 1785. By 1810, output per head was at three-quarters of the British level, and much of it was exported to the three neighbouring countries. By 1831, almost 0,5 m tons of the total output of 2,3 m t was exported, and by 1861 it was 3,4 m t out of 10 m t. As in Britain, much of Belgium's industrial geography was determined by the coalfields. (Wrigley 1961, Pounds 1985, Mitchell 1981, Lebrun 1979. Also Table V.)

Iron ore was found close to the coal and Belgium remained self-sufficient until 1860. The iron-working industries around Liège were among the oldest in Europe making use of the plentiful local water power. Early in the nineteenth century Belgian iron-masters turned to modern methods and were responsible for several technical innovations themselves. In the 1820's and 1830's British coke smelting and puddling technologies were widely adopted, among the first on the continent (Craeybeckx 1970, Dhont and Bruwier 1973, Lebrun 1979, Pollard 1981, Pounds 1985). Some of the earliest continental engineering works were erected in the Liège region and in Ghent, (Cockerill's of Seraing being for a time much the largest in Europe) and supplied not only Belgium, but also Germany, Russia, the Netherlands and other parts of Europe, as a kind of indirect source of supply of the latest British technology (Henderson 1954, Chapter 5, Dhont and Bruwier 1973, Mahaim 1905).

The woollen industry had a long tradition at Verviers and was not far behind the British in modernizing itself. Ghent had a flourishing calico printing industry in the eighteenth century. One of the earliest continental modern cotton spinning mills was established there in 1801. Though the cotton industry had its ups and downs, it survived and expanded in the long run. Only the traditional Flemish linen industry failed to switch to modern methods, though some mechanised mills were established in other parts of the country. Other industries based on coal also did well, such as chemicals, and in glass-making Belgium for a time took the lead in Europe.

The southern Netherlands had been early in the field in the building of an efficient canal system, which particularly benefited the coal mining industry. In the 1830's Belgium became the first continental country to build a substantial railway network. The first major lines were laid down by the State from 1834 onward, but after 1850 the system was built out by private companies (Clapham 1963). Belgian engineering and iron works supplied locomotives and rails, respectively, to other countries in considerable quantities in mid-century. By 1860 it may be said that the first decisive stage of Belgian industrialization had been completed.

Policies and Causes
Belgium became formally part of revolutionary France in 1794 and, being the most developed of the French provinces, benefited greatly from Napoleon's policy of furthering French industry, while enjoying free access to the large French internal market. There followed the union with the northern Netherlands in 1815–30, when she again benefited from the deliberate efforts by the Dutch king to foster the industries of the southern part of his kingdom. After independence, she had a government not dominated by powerful agrarian interests, almost unique in Europe, which continued to offer much official support.

Among the more important measures of State support, beside the railway building noted above, were a protectionist policy as long as Belgian industry needed it, followed by trade agreements with the large neighbours to ease Belgian industrial exports there, and a series of laws which hampered trade unionism and favoured employers. The formation of joint-stock enterprise was made easy, and Belgium was distinguished by the early establishment of a major bank, the Société Générale, set up with the help of royal funds as early as 1822, to help the development of industry. This was followed by other industrial investment banks which made capital available especially for the heavy industries (Milward and Saul 1973, Clapham 1963).

Being a small, largely industrial country without a politically powerful landed nobility was thus one of the advantages enjoyed by Belgium to explain its early start. Another was its long urban and industrial tradition, among the earliest in Europe, and its highly developed and efficient agriculture. Peasants on holdings too small to afford a living so that the rural population had to seek at least part-time work in industry (Mendels 1972) helped to keep wages low, and low wages were certainly among the major causes of Belgian precocity (Mokyr 1976). The location between the most advanced countries of Europe helped to encourage an early railway network, and proximity to England made it easy to copy British techniques as well as attracting British specialists who founded several of Belgium's early firms, and British capital to help build her railways. Above all, however, her natural resources, water power and water ways, coal and iron ore, must be counted among the major causes of her early industrialization.

Political changes did not always work in Belgium's favour. The sudden loss of access to the French market in 1814 to which her industry had been accustomed, and still more the separation from the Netherlands in the revolution of 1830 caused severe shocks, which explain in part the fitful and irregular course of Belgian economic growth (Dhont and Bruwier 1973).

Here lies a major difference with the British industrial revolution which took place within unchanging borders, except for the incorporation of Ireland into the United Kingdom in 1801. A further difference was the greater dependence on exports to her immediate neighbours because of the small size of her internal market, and the ability to take her technology ready-made, together with experienced managers and workers, from the United Kingdom.[10] Conversely, the State played a more active role than in Britain, and so did the large joint-stock banks, both of which may be connected with Belgium's role as follower rather than leader. If a multitude of protestant beliefs is said to have furthered British industrialization, the Roman Catholic religion did not hinder that of Belgium. In other respects, however, the Belgian industrialization was probably the process which followed the British model most closely on the continent.

1.3. Switzerland

Switzerland and Belgium are often treated together, as two small countries which led the industrialization of the continent (Milward and Saul 1973). Their position in relation to the other early industrial countries is illustrated in Table IV. Switzerland, though a smaller country, fully kept pace with Belgium well ahead of the larger countries, on the basis of a "level of industrialization" adopted by Bairoch as a standardised method of comparison (also see Table V). Yet most of the favourable pre-conditions enumerated for Belgium were missing. Switzerland has virtually no coal, iron or other mineral deposits. It is landlocked, it has the highest mountains in Europe blocking access to neighbours in all directions except the North, and the neighbours, additionally, mostly maintained high tariffs against Swiss imports. Until 1848 there was no Government, properly speaking: each canton governed itself and levied excise duties on the products of other Swiss cantons on the same basis as on foreign wares. Even after 1848 Switzerland kept to a virtual free-trade policy.

There were factors which worked in favour of the Swiss but some of them require an explanation in their turn, such as the high level of eighteenth-century industrialization on the basis of textiles and watches, and the large capital accumulations in the cities. How could these have come about in a country so ill-favoured with resources? Why did Switzerland not resemble the less accessible Balkan regions, but became a leader among European industrial nations? The Swiss example strongly suggests that the significance of some of the factors favouring that country is often underrated. Among these was freedom from a parasitic landed nobility and from its power to legislate in favour or agrarian interests; social mobility and urban self-government; the Calvinist–Protestant religion and the high level of education achieved; and associated with all of these, an active, innovative class of entrepreneurs, managers and engineers.

Given the resources of the country, industrialization naturally took quite a different path from the Belgian. Instead of coal, crude iron and cheap cottons, the Swiss had even in the eighteenth century concentrated on high quality silks, printed cottons and watches. The first mechanical cotton spinning mill was erected in 1801, and chemical bleaching tried out in 1801–2, but despite the use of British-built machinery and the presence of some British engineers, the Swiss did not follow the British into the market for cheap mass production. To

fine yarns and weaving was soon added embroidery, for which inge-
nious machinery was invented locally. The number of cotton spindles
rose from 400,000 in 1827 to 920,000 in 1853 and almost 1½ million in
1864, the tonnage of yarn spun from 680 in 1814 to 2,800 in 1827 and
8,000 around 1850 (Bergier 1983, Pollard 1981). In cotton consumption per head, the Swiss were far ahead of other continental countries
(Table V). Silk weaving and watch manufacture, which continues to
flourish, produced goods of a high value-to-weight ratio able to carry
high transport costs.

Swiss engineering ingenuity was most remarkable. Possibly the
precision work required in watch making may have contributed to this.
Steam power came late, but it was in textile machinery that the early
engineers showed their skill and innovative abilities, while J.C. Fischer
pioneered the production of cast steel in Europe (Henderson 1954,
chapter 9, Lincke 1910). By the 1850's, machinery exports greatly
exceeded imports. For railways the terrain was difficult, and little was
done until the 1850's. But then building proceeded apace (Table V), in
part with the help of British engineers and French capital.

After 1848 the internal duties fell, and a common, if low tariff
against imports raised some revenue for the Government, but there was
little State initiative and a maximum of *laissez-faire*. Employment in
industry had risen to 33% of the employed population in 1850, and to
42% in 1880. Factory workers proper increased from 50,000 in 1850 to
134,000 in 1880/2 (Bergier 1983, Milward and Saul 1973). At the same

TABLE V
Indices of Industrial Development, 1840–1860[a]

	Raw cotton consumption, kg per head		Railway development index[b]		Coal consumption, kg per head		Fixed steam power, HP per 1,000	
	1840	1860	1840	1860	1840	1860	1840	1860
Belgium	2.8	2.9	6.6	30	850	1310	8	21
Switzerland	3.7	5.3	negl	28				
France	1.5	2.7	1.2	18	130	390	1	5
Germany	0.9	1.4	1.1	21	110	400	0.6	5
United Kingdom	7.3	15.1	7.2	44	1110	2450	13	24

[a] Source: Bairoch (1965) pp. 1102, 1106–8.
[b] The index is a compound of railways per head and per square kilometre.

time, the proportion of food imports rose from 20% of consumption around 1850 to around 50% at the turn of the century. The tariffs of neighbouring countries were by-passed by exports to the USA, to Latin America, the Levant and even the Far East.

Bergier's (1983) "Swiss Model" of industrialization starts with high grade cotton and other consumption goods, makes use of highly skilled, yet still cheap labour, in relativery small factories in rural setting which are able to use water power. Where possible, these would be supplied by components made by domestic labour from the surrounding countryside, and when that became scarce, also from neighbouring territories outside the Swiss border. Pre-existing merchant capital helped in trading, but little outside capital was required in industry.

It is a model very different from that of Belgium or of Britain. In its emphasis on high-grade fashion goods and on the skills necessary for them, it bears some resemblance to the French model. France, however, constitutes a type of her own.

1.4. France – An Alternative Type?

Slow Growth and a Choice of Models

There is general agreement that French industrialization took place against a background of very slow economic growth — but this, as in the British case, is to be understood in relation to later industrialisers, not to earlier centuries. Thanks to a massive research effort extending over many years, precise figures are now available to document the growth of the physical product of agriculture and of industry, as well as of national income in the nineteenth century. For the period 1815/24 to 1905/13 Marczewski (1965) calculated an average industrial growth rate of 2.0% p.a., which, when modified by the low agricultural growth rate of 0.7%, averaged at 1.5% a year for the physical product. Markovitch (1965) found all industry to have grown by 2.03% a year over 1815–1914, but if certain slow growing sectors were excluded, the average rose to 2.84%. Lévy–Leboyer (1968) calculated 2.20% for all industries, 2.56% for the faster-growing ones. Crouzet (1970, 1972) reached 2.97% for the bulk of the industrial sector, but this was reduced to 1.8% if the slow growers, linens and woollens, were included. Fohlen (1971) also opted for 1.8%.

Taking the French level of physical output as a whole, its growth rate of 1.5% a year lay well below the British rate of 2.65% for this period; but given the slower French population growth, output per head, at 1.2%, was virtually identical with the British ratio of 1.3%. The same may be said about the French national income growth of 1.5% a year per capita between 1825 and 1874 (Lévy–Leboyer 1985, Caron 1979, Cameron and Freeman 1983).

Modern French economic growth (like the British) began early in the 18th century, following a disastrous period of decline c. 1680–1720. Up to 1789 output, as well as overseas trade, kept pace with the British (Leet and Shaw 1978, Crouzet 1967, Léon 1960, 1974) and did so once more after the break of the revolutionary and Napoleonic wars. What is not clear is the absolute distance between them, or, in other words, the number of years by which the French were trailing behind Great Britain. Estimates vary between close similarity and up to 20% difference. Bairoch's calculations, (Tables IV and V above) show a wider gap; his estimates for national incomes per head (Table VI) put France in the years under discussion at only around two-thirds of the British level, but very close to the other advanced western European countries.

France possessed a very different range of resources, her political history diverged widely from the British, and she had to contend, as Britain had not, with a powerful more developed neighbour. Nevertheless, the similarities were close, and France never lost contact with the most advanced British technology. French industrialization is therefore, as a rule, seen in relation to the British. Given the similarities and the differences, three models or patterns of observation have emerged

TABLE VI
GNP per Capita, Western European Countries, 1830–1860[a]

	1830	1860
United Kingdom	370	600
France	275	380
Belgium	240	400
Germany	240	345
Switzerland	240	415
Netherlands	270	410

[a] Source: Bairoch 1981, p. 10. The sums are expressed in U.S.$ of 1960.

to describe the French industrialization process, though the dividing lines between them are somewhat blurred.

(1) Some historians have seen the French industrial revolution as a copy of the British, trailing by some decades. This has raised questions as to the French "failure" or as to the causes of retardation. (2) A second view has been to take the British experience as the model, but to stress the deviations made necessary by differences in resources and markets. This allows the French some "success" in adapting to circumstances. (3) A third view denies the relevance of the British model altogether, and sees the French path as an altogether different one, albeit ultimately aiming for the same target.

One extreme version of the first view treats the position of the two countries, and the French chances at the outset as equal, and sees the start in Britain rather than France as accidental (Crafts 1977). For Rostow (1978), France followed the same set path, her "take-off" period being 1830–1860, while the increase in her investment ratio fitted the Rostow–Lewis thesis rather better than the British, net capital formation as percentage of net domestic product rising neatly from 3.0% in 1788–1839, to 8.0% in 1839–52 and to 12.1% in 1852–80, the gross figures being 6%, 11% and 17% respectively (Rostow 1978, p. 399, Marczewski 1963, Lévy–Leboyer 1978). Crouzet saw France as one of the western countries which were " 'followers' or 'latecomers' who tried, during this period (sc. 1815–1850) to emulate Britain and to introduce the new technology which she had invented".[11] There is much overlap, but no agreement (any more than there is in Britain) when exactly the French "take-off" occurred. Crouzet saw 1815–50 as the "Gründerzeit" (1972c, p. 108), Dunham put it at 1815–48, with maturity by 1860, Fohlen (1970) and Trebilcock (1981) marked out 1850–70, Pierre Léon (1960, p. 185) noted a "grande accélération" in 1830–75 and "grande rupture" in 1840–70 (in Braudel and Labrousse 1976), for Palmade (1972) the "great upsurge" was dated 1848–1882, while Roger Price (1981) is inclined to let the transformation begin in the 1840's, and Kemp (1971, p. 110) agreed that 1815–48 were "years of preparation rather than achievement". The industrial sector showed a clear, if limited acceleration in 1815–1845 (Table VII).

Many different causes have been made responsible for the retardation, or "backwardness" (Gerschenkron 1966b) of the industrialization process in the French economy. In the literature of the first postwar decades, lack of entrepreneurship was assigned much of the blame:

TABLE VII
French Industrial Growth, 1781–1874[a]

	1781/90–1803–12	Aveage Annual Growth, per cent		
		1803/12–1815/24	1815/24–1845/54	1845/54–1865/74
23 Industries	0.5	2.4	2.46	1.97
19 Industries[b]	0.6	2.2	2.63	2.43
19 Industries Altern.	1.1	2.4	3.00	2.98

[a] Source: Markovitch (1965) p. 38.
[b] The four industrial groups omitted were: construction and public works, food industries, clothing, wood and furniture.

French industrialists, it was said, were unambitious, family-oriented, non-competitive and conservative (Landes 1969, Lévy-Leboyer 1976, Fohlen 1978, Price 1981). This is now largely discounted as many different types and qualities of French entrepreneurship have been found to have existed. There is broad agreement on the relatively poor resource endowment: France has little coal, the coalfields and iron ore fields are widely scattered, and the geographical location of such resources as there is often difficult of access (Pounds 1985). Political upheavals and costly wars are sometimes given some blame (Fohlen 1971), as is the scattered market at home in view of the small urban sector and the poverty of the agricultural community, and the limited markets abroad because of the predominance of British exporters. The backwardness of French agriculture has been widely seen as a major retardative factor. Lastly, and more rarely, the cause of French backwardness has been seen as a failure to invest (Cameron 1958). Several of these points will be discussed further below.

It is France's different resources, institutions and traditions, as well as the existence of a powerful advanced economy across the channel, which the second group of historians, those who see French industrialization as a distinct variant of the British, tend to emphasise. "Industrialization in the French style"[12] concentrated on high-quality and luxury goods, making use of traditional skills and artistic sense, of cheaper labour and, not least, of the survival of Paris as a centre of fashion from the days when France was still the leading economic power in Europe. Mechanisation tended to go "upstream", beginning with the finishing sections and the consumption industries sections in contrast to Britain's "downstream" mechanisation beginning with the early processes and capital goods (Lévy-Leboyer 1968, 1976, Roehl 1976, Cameron and Freedeman 1983, Crouzet 1977c). In this way France filled intelligently the gaps left by British dominance in mass-production industries, and the two countries become complementary rather than competitive. While Britain sent mostly raw materials (coal) and semi-manufactures to France, France exported finished manufactures and food (including wines and brandies) across the channel. If French firms and factories remained smaller, that, as a recent econometric study seemed to show (Nye 1987) need not have been a disadvantage.

The third school of historians considers the differences between the two countries to be large enough to describe French industrialization as of a different type altogether. In an extreme form, it has even been said that it was France which, according to on the Gerschenkron concept, was the first industrializer, Britain being the more "backward" economy (Roehl 1976). This view has found little support, since it is, according to that concept, not merely necessary for the pioneer to grow slowly, but to reach the breakthrough to modern industry first, which cannot be said of France. However, France as the "most aberrant case"[13], as the country, moreover, that more clearly became a model for the European latecomers than did Britain, has found more widespread support (Cameron 1985, Cameron and Freedeman 1983). "Emulation of British technology and practices across a wide range of industrial activity", according to O'Brien and Keyder[14] "seems irrelevant to the needs of France for most of the nineteenth century. In any case, what France actually did in industry seems to have been accomplished, for the most part, as efficiently as could be expected".

In France, no "take-off" could be discerned (Marczewski 1963, Lévy–Leboyer 1968), her growth started very early, but was more limited in scope, with a poorer showing in agriculture and the tertiary sector, it reached higher levels of income before changing the nature of industries, required less urbanisation, and less division of labour with the rest of the world. It caused less upheaval in social life (though it was accompanied by more political disruption), and according to some, it almost came by stealth. The term "industrial revolution" applied to France, at one time quite common (e.g. Fohlen 1970, 1973, Henderson 1967, Crafts 1977) has become quite rare.

Structure and Industrialization

French industrialization is widely considered to have been hampered by a backward agriculture. Though there were some regions with large, efficient farms, such as the North and the Paris basin, and French farmers reacted quite well to market changes, the bulk of the land was worked by a poor, badly educated, conservative peasantry, holding their land firmly after the revolutionary settlement which abolished all feudal dues and, unlike the peasantry in the rest of Europe, gave the peasant possession without obliging him to pay off the former owners (Milward and Saul 1973, Pollard 1981, Caron 1979, Price 1981). French agriculture even in 1840 has been called "stagnant, backward and primitive".[15] The drag on the economy was made worse by the large size of the agricultural sector: it accounted for 57% of commodity production in 1781-90, and 47% still in 1865-74, while its share of the labour force in the same two periods fell only from 81% to 65% of those engaged in commodity production.[16] However, not all judgments are negative. Agricultural output increased over the century by about 1.1% a year, a maximum rate of 1.39% being reached in the decade 1855-64, and cereal output grew even faster, at over 2% a year.[17] After the mid-century, France, aided by her growing railway network, was able to escape the cycle of recurring famine. More capital began to be employed in the form of better equipment and more fertilizer, and methods were improved (Price 1983, Braudel and Labrousse 1976, pp. 987 ff.). As soon as conditions became favourable, therefore, the agricultural sector proved that it could supply more food as well as labour and capital to the rest of the economy (Newell 1973, Heywood 1981, Lévy-Leboyer and Bourguignon 1985). Yet it could not be denied that the French agricultural transformation did not precede the industrial transformation, as in Britain, but was at best simultaneous with it.

The thesis of O'Brien and Keyder (1978) that French inferiority in agricultural productivity, compared with the British, was matched by a considerable superiority over British industry has not found much support, for it would imply that each country concentrated on the sector in which it suffered comparative disadvantage; moreover, the explanation that the French, who had cheaper labour, achieved better results in industry by using more capital would also imply that each country used more of that factor of production with which it was badly provided, in direct contradiction to the assumptions of the Heckscher-Ohlin theory. However, there is wide acceptance of their more general point, that it was the preference of the Frenchman to stay on the land and his ability to do so, even if under-employed on his peasant plot, compared with the high mibility of the British landless labourer, which accounted for the persistence of a large gap in France between the low productivity and low incomes on the land compared with industry. Urban industrialists, in turn, deprived of full-time labour, were tempted by low wages in the countryside to persist with technically backward putting-out systems instead of converting to factory organisation (Lévy-Leboyer 1976, Price 1983, Nardinelli 1988).

France, as much the largest economy in Europe at the beginning of the nineteenth century, possessed an extremely wide spectrum of

industrial plants. There was scarcely an industry which was not represented somewhere in the country but, influenced by differing factor
supplies and frequently protected until the railway era by distance,
firms varied much in efficiency and productivity.

Among textiles, linen as well as the woollen industries of southern
France stood up least well to British competition. In the north, a
mechanised woollen industry did develop, though technically well
behind the British counterpart, while Alsace, in spite of its unfavour-
able location, became a leader in high-class cotton goods, and the
North developed a mass-production cotton industry. The silk industry,
centred on Lyons, kept in the technical forefront, adopted refined
systems of division of labour and continued to dominate European
markets. The Parisian fashion industries, unregenerate in their lack of
mechanisation and factory organisation, remained a pillar of French
exports.

It was among the mining, metallurgy and engineering sectors that
France trailed most seriously. Coal was dear and its supply scattered,
some had to be imported, and the Pas de Calais field, rich but difficult
of access, was discovered only in 1846. In the traditional charcoal iron
industry, French output had long exceeded that of Britain, producing
130–140.000 t in c. 1788 against the British pig iron output 63.000 t
(Crouzet 1967), but the new technologies were adopted only after much
delay. Only in 1818 were the first coke blast furnaces erected in Le
Creusot and Pont L'Evêque, the first puddling hearths followed in
1820, almost forty years after the British, and as late as 1824, there
were only six steam engines employed in the whole of the French iron
industry (Vial 1967), though by 1859 there were 1040. The technical
know-how and much of the initiative for the pioneer installations had
came from British entrepreneurs, managers and workmen as well as
from Frenchmen who had gone to Britain to learn — as indeed was the
case in some of the textile trades; even so, progress was slow, and until
well into the 1850's, charcoal pig exceeded the tonnage produced in
coke furnaces. Around 1860 French pig iron output of 25 kg per capita
compared with 130 kg for the United Kingdom (Bairoch 1965, p. 1104).
Different judgments are possible about this record. The delay may be
seen as a consequence of French conservatism or incompetence; but in
view of the cheapness of charcoal and the dearness of coal, it may also
be seen as a sensible adaptation to circumstances.

The same may be said about engineering. French steam power, in spite of the work of British emigrés, was slow to develop (Table V), but great strides were made in improving water wheels and turbines, and later France was to lead in hydroelectric technology (Fohlen 1973, Henderson 1954, Part. II). France had 10.000 HP installed in 1830, 60.000 HP in 1848, 205.000 in 1862. While total industrial growth in 1815–1913, was at the rate of 1.61% a year according to Crouzet, mining grew at 4.2%, primary metallurgy at 3.5%, and chemicals at 4.2% (Crouzet 1970). Similar accounts, of occasional high achievement, but general backwardness, a "dual industry structure", could be given for other industries, such as glassmaking, leather, paper, or gas lighting (Lévy-Leboyer 1968).

Much was spent by the July monarchy on the improvement of roads and canals, and indeed as late as 1855–64, average annual goods traffic amounted to 2.7 milliard ton/km by road and 2.1 milliard ton/km by canal and sea, compared with 3.0 milliard ton/km by rail (Price 1981, p. 36). But it was the railways which provided the dynamic element to the French economy in mid-century, in their backward linkages to the iron and engineering industries, in their forward linkages by ending at last the isolation of the French provinces, and by their influence on the methods of mobilising French savings (O'Brien 1983).

Railway development was slow at first. By 1842, only some 600 km were open (against 3.100 km in Britain), an important part of which, the Paris-Rouen-Le Havre line, being largely built with the help of British capital, British engineers, and British labourers, even the attached locomotive works being largely British (Henderson 1967). In that year, a national plan was agreed, the State to decide on the routes and to lay down the permanent way, and private companies to build the superstructure and run the lines. By 1848 2.000 km had been built, but many stretches were half-finished and the companies were unable to complete them when the crisis struck. Under the Second Empire a new form of concession encouraged renewed building, and, following the next crisis of 1857, the State undertook to guarantee interest payments to companies under the "Franqueville conventions" to ensure that first a system of main routes, and later of local routes was completed; companies were amalgamated into six main networks, each covering a segment of the country. By 1870, 17.400 km of line were in operation. The main effort, therefore, had come after 1850 (Milward and Saul 1973, O'Brien 1983, Braudel and Labrousse 1976).

By 1847, 1,638 million francs had been spent on the building of the railways; in the peak period 1853–1867, annual investment in the railways ran at the rate of 393 million francs (Caron 1970). To raise these sums, new organisational forms had to be created. The formation of joint-stock companies with limited liability was simplified by the legislation of 1863 and 1867. Above all, new types of banks were founded. Up to then, the banking system had consisted, apart from the Banque de France (1800) as central bank, conducting Government business and issuing notes, of private merchant bankers mainly dealing with state loans and foreign trade, of "departmental" banks with restricted functions, and of small local banking houses for short term loans. The merchant banks also occasionally helped very large companies with raising long-term capital. The smaller saver had put his money in land or urban property, or kept it in cash. He was now to be attracted by large privileged companies, of which the *Crédit Foncier* (1852), and the *Crédit Mobilier* (1852) were the prototypes. The latter especially, in competition with some of the older merchant banks such as Rothschild, helped to raise much of the capital for railway building both in France and abroad, though it did it by the risky method of borrowing short and lending long which led it into trouble in the crisis of 1867. A second type of bank also emerged, attracting deposits from the public and investing them in gilt-edged papers. This type, beginning with the *Crédit Industriel et Commercial* (1859) and the *Société de Depots et de Comptes Courants* (1863) as well as the *Crédit Lyonnais* (1863) also developed branches and provided some credit for industry (Palmade 1972, Braudel and Labrousse 1976).

The French structure resembled the British inasmuch as the banking system provided no capital for industry in the industrialization phase, except in the case of a few large joint-stock companies. It differed in not providing much current credit, either; indeed, before the 1860's there was little paper money available to eke out the money stock beyond the gold and silver coins. Thus a dual system of large, well-capitalised modern firms, and small stunted firms was maintained, the "lack of easy access to short-term credit" condemning the latter "to remain small, and eventually bred the psychological outlook associated with petty capitalism".[18] Cameron (1967) may be somewhat unjust in blaming the slow progress of French industry, which in the 1830's and 1840's was allegedly "straining for its 'take-off' "[19], on the lack of short-term credit and the shortage of means of payment (Plessis 1987). But it may well be that inadequate institutions and the abuse of its monopoly by the Banque de France slowed some developments before the 1860's.

The Role of the State

The French revolutionary governments are generally credited with having carried through a radical transformation of French society and the French economy. They are also said to have driven the landed nobility from power and ushered in an era of dominance by the bourgeoisie.

A great many changes were indeed carried through in the years after 1789. The church and most emigré aristocrats lost their land, and the peasant, as noted above, often became a small proprietor. Gilds were swept away in 1791, internal tolls abolished 1790, the existing commercial treaties were denounced in 1792, soon to be followed by tariffs and a Navigation Act. A Patent Act was passed in 1791 and new weights, measures and a metric coinage laid down in 1795. Trade unions were prohibited in 1791 and the compulsory *livret* to be carried by workmen from 1803 further restricted the freedom of the wage owner in favour of the employer. There were tax reforms and efforts to improve technical education at all levels. Napoleon carried on with similar policies, his Civil Code of 1804 being built on a concept of private property which corresponded to that of the capital-owning classes. The Companies Act of 1807, while it made limited liability difficult to obtain, facilitated the formation of *sociétés en commandite* — the form taken by most larger enterprises in the following decades. Further, the continental system, while mainly designed to bring Britain economically to her knees, provided protection for French entrepreneurs and led to much industrial growth, though it also isolated the French further from the latest advances of British technology (Kemp 1971, Henderson 1967).

All these, doubtless, favoured the rising bourgeoisie. But it would be wrong to assume that the landed nobility had lost all their property and power; they kept both until well into the second half of the nineteenth century (Beck and Beck 1987). At the same time the middle classes themselves were attracted more to conservative types of property, including rural and urban land, and the traditional professions, than were the British (Kemp 1971). Thus they were not in a strong position to exact financial support from the State, except by the indirect method of the enormous sums paid out in one form or another for the railways.

These, by definition, went mainly to lines which carried little industrial traffic: those that did, needed few subsidies.

The main Government support for French industry came through the tight system of protection. France was the classic protectionist country in the nineteenth century. The fact that the system was (temporarily) dismantled in 1860 may be taken as a sign of French economic strength and self-confidence, though free trade had to be enforced on a bitterly resentful business community. Only some of the successful export industries, like silk luxury products, Alsace cotton and wine, were fully in favour.

However, even before 1860 many duties, especially on raw materials, had been reduced: the average customs duties on imported goods had declined from 17.2% of total value of imports in 1847-9 to 10.9% in 1855-9, to sink further to 6.2% in the first years of the commercial treaty system inaugurated by the treaty with the United Kingdom in 1860.[20]

Under the latter, all prohibitions were lifted, and the maximum French tariff was fixed at 30% ad valorem, to be reduced ultimately to 25%. In practice most were much below this, while Britain freed practically all imports from duty. The effects on the French economy of this treaty, and of others with Belgium, the Zollverein and other countries which followed in quick succession are much debated, which points to the fact that they could not have been very clear-cut in one direction or the other. For many firms, it provided a necessary cold douche of competition to rouse them into a competitive stance, helped in some cases by payments out of a special Government fund. Some others, especially in textiles, as well as in shipping after the navigational subsidies had been reduced, no doubt suffered, but it is difficult to disentangle the consequences of the freer trade from other changes of the time (Rist 1970).

It may be that French taxes bore more heavily on French firms than on British (Cameron 1958), but underneath all the superficial distinctions between the classic *laissez-faire* economy of Britain and the classic *dirigisme* of France, the differences tended to disappear in the phase of industrialization. The restrictiveness of the gilds in eighteenth-century France has often been exaggerated (Crouzet 1967); both countries maintained protection in the early phases, to turn to freer trade thereafter (though France quickly reverted to protection again); both were relatively backward in educational provision, both were eager to encourage company enterprise with a minimal protection of the public, making some provision for patents for inventors, and in both, with all their different traditions, the law was adapted to a capitalistic interpretation of private property. In the end, it was only the revolutions and the costly wars, inasfar as they were the results of State activity, by which the State may be said to have prevented the French economy from fully keeping up with the British (Leet and Shaw 1978).

France's Role in the World
In the eighteenth century, France was a major colonial power, acting as an intermediary for colonial produce for other continental countries. Just over a third of her exports, a ratio which did not change over the century, consisted of manufactures, though few of these went to the advanced countries, Britain and the Netherlands (Léon 1974). Much of the empire was lost in the wars against Great Britain, and in the decades of industrialization, colonial trade played a minor part in the French economy. In the war years the British Orders in Council severely damaged French overseas trade, but it expanded remarkably quickly again thereafter. Taking 1825–34 as 100, exports of manufactures rose to 257 in 1845–54, and to 607 in 1865–74, while exports of food and raw materials rose to 213 and 1000. Similarly, imports of manufactured products rose to 144 and 906, and of food and raw materials to 200 and 674 between the same years. As late as 1860, France was still the second trading nation in Europe, holding 19.2% of European trade against 29.8% for Britain and 18.4% for Germany. As a proportion of GNP, French trade rose from 13% in 1830 to 41% in 1870. With both Britain and the USA France had a healthy export surplus, mainly in goods of high quality and high unit value. Between two-third and three-quarters of French trade continued to be done with Europe (Lévy–Leboyer and Bourguignon 1985, Braudel and Labrousse 1976).

As the largest and most advanced European economy, France had been a major source of science and technology in the eighteenth century. In the period of industrialization, much of this continued, in spite of the growing British technical lead over France. French engineers, in particular, helped to develop mines and railways as French capitalists supplied capital for railway building abroad (Cameron 1961).

Substantial foreign investment began after 1850, when French foreign holdings are estimated to have amounted to around 2 milliard francs. About 40% of this sum was invested in Spain and Portugal, another 40% in Belgium and Italy, and virtually all the rest in other parts of Europe. By 1860 the total had reached 10 milliard (as against 2,3 milliard of foreign money invested in France) and by 1880, after the disaster of the lost war against Germany and the indemnity, it stood at 15 milliard (against 4,5 milliard foreign money in France). In the 1850's, some 2.7% of GNP was invested abroad every year, and in the 1860's this had risen to 3.4%.[22] In addition to railways, much of it was in public loans (White 1933, Feis 1964, Lévy–Leboyer 1977).

In the eyes of some, these investments were not in the French interest; the country would have done better, in that view, had these sums been invested at home to benefit French industry or agriculture. On the other hand, French investment at home were at a respectable level; it was merely French savings that were unusually high in the Europe of the day (Milward and Saul 1977). It was in the nature of French investors that they should prefer the steady, if not always safe, returns of foreign Governments or railways to the fluctuating returns of home industry. In much of the rest of Europe, railways could be built ahead of local capacity (and often ahead of economic need) because of the flow of French surpluses. They speeded these countries' development and linked them earlier than would have been the case otherwise to the industrialized parts of the world.

2. RAILWAYS AND COMPETITION: GERMANY AND THE USA

2.1. Germany, the Model Follower-on

Growth and Transformation
Germany is usually classified among the second group of followers-on, a generation behind the first line of countries discussed so far. She is also held to be a model pupil, the most successful industrialiser on the European continent, in stark contrast to the alleged failures of the French economy.

It is not clear why Germany should be considered to have lagged much behind France. A glance at Tables IV, V and VI will show that she was either just behind, equal or even (in railway building) ahead of her western neighbour. Her "take-off" is said to have occurred about the same time as the French. Rostow, who was originally inclined to date it 1850–73, revised it to 1840–70 following Hoffman, who even put it at 1830/5–1855/60 (Rostow 1978, pp. 401, 407) or alternatively dated its start in the 1840's (Hoffman 1963). Henning (1984) considered that Germany's "preparatory phase" was over by 1835, although Borchardt (1973) thought that the "preliminary phase" ended only in 1850. According to Tilly (1978), the "decisive breakthrough" occurred in the 1840's, though it was halted in the sharp crisis of 1845–7, to take off properly in 1850–73. Treblicock (1981) also places the "take-off", somewhat doubtfully in 1850–70. Mottek (1964) put the beginning of the industrial revolution in the late 1830's. Clapham (1963), writing before Rostow's stage theory came into vogue, considered the acceleration to have been "perceptible from about 1835, and conspicuous from about 1845".[23]

These dates are not very different from the French, the main distinction being that there seems to be less doubt that a "take-off" had occurred at all. Curiously, however, precisely this distinction from the French is not borne out very clearly by the available statistics.

According to Hoffman (1965, p. 20) the net domestic product of Germany in the period 1850–1913 rose by 2.6% a year and the numbers employed increased by 1.2%, which left output per employed person to have risen by 1.5% — very close to the French rate. Because of reductions in hours, output per hour of work, however, rose by 1.8% a year.[24] On Kuznets' (1971b) different base, the German decadal growth of 16.3% per head in 1850/9 to 1910/13, is, if anything, rather lower than the French rate of 21.6% for 1831/40–1861/70, and of 13.5% for 1861/70–1891/1900, which averages out at around 17.5%. Nor is the German growth of 16.4% per decade for the 1850's to 1880's, and 17.0% for the 1880's to 1905/13 very different from the British of 20.4% and 11.4% for similar periods.[25] Average incomes per head in constant marks of 1913 rose from 265 in 1850 to 593 in 1900 (Henning 1984) or, according to Hoffmann (1959, p. 14) from 295 in 1850/5 to 728 in 1911–13. A difference of 0.1% or 0.2% a year can be significant if continued over the large part of a century, yet the very different treatment accorded the two countries in the literature remains somewhat puzzling (Kaelble 1983).

One reason for it may have been the apparently lower starting level of Germany, which up to then had appeared as a sleepy hollow with just a few industrial concentrations in areas like Saxony, Upper Silesia and Rhineland-Westphalia, compared with the powerful French economy which had dominated Europe for so long. Moreover, while the per capita growth rates, which are in one sense the fairest measure of progress, show no great difference, German total growth was indeed much faster than the French, given the rapid population increase. The absolute size of the country also made an impact on contemporary consciousness and helped to make Germany the dominant economic power on the continent by the end of the century.

If we bear additionally in mind the relative political impotence of the German mini-states up to then, it seemed indeed as though "all the forces tending towards industrialism and urbanisation had struck Germany at once."[26] Transformation at many levels, including the unification of the country as a consequence of three victorious wars in quick succession emphasised that impression.

The rate of capital investment, for example, bore a strong likeness to the revolutionary change predicted by the Lewis-Rostow theory. Estimated at 5% of national income before 1850, with a rising tendency and a strong concentration on the traditional sectors of agriculture and building (Tilly 1978, p. 427), net investment as a proportion of net national product shot up to 9% in the 1850's, 12.5% in the 1870's and 13.3% in the 1890's at current prices; the percentages at 1913 prices were 7.3%, 11.0% and 13.9% respectively (Hoffman 1963, 1965, pp. 104-5, Borchard 1973). The capital stock increased fivefold between 1850 and 1913 and its growth accelerated from 2.3% a year in 1850-75 to 2.7% in 1875-95 and 3.4% in 1896-1913; in this, the share of industry, dwellings and railways increased strongly at the expense of agriculture and roads.[27] The share of capital goods industries expanded correspondingly within total industrial growth.

This concentration on the heavy industries (in contrast with the French weight on textiles), as well as the widespread employment of banking capital in industrial investment, another feature of the German industrial revolution, fits into the Gerschenkron (1966) pattern of the later developer, as does the more rapid "spurt". Moreover, within a similar overall growth rate in Germany as in France, there were yet several indicators in Germany pointing to a course which had a great deal in common with the first industrial revolution in Britain. One reason for this must lie in the resource endowment — especially coal and iron ore — and was reflected in the nature of the structural transformation of the German economy.

Structural Change

Agriculture reflects, perhaps more than any other sector, the wide span of regional variations of the German economy, stretching as it does from the south-western and western small-farm regions which showed strong similarities to their advanced neighbours in France and the Netherlands, respectively, to the East-Elbian knightly estates ("Rittergüter") which approximated to the most backward feudal agrarian regions of Eastern Europe. It was to be highly significant for the industrialization process and the general development of modern Germany that the farmer-landlords, the Junkers, of the eastern provinces of Prussia came to dominate the country politically and, after the unification of Germany under Bismarck, extended their power over the whole of an industrialised Reich.

Agrarian servitude of an almost classic feudal form prevailed in the East-Elbian provinces until 1807. The October emancipation edict of that year was passed in an upsurge of national feeling after the devastating military defeat at the hand of the French — as well as the need to pay contributions to Napoleon — had made reform inevitable. It provided personal freedom at once, but left the distribution of property until later. When that was taken in hand, beginning with the regulations of 1811 and 1816, the Junkers had recovered most of their power and the details of peasant emancipation were settled very much in their interest. Above all, it was assumed that the land and the peasants had been their property, so that they had to be "compensated" for their loss. Peasants who had hereditary rights had to hand over one-third of their land in exchange for their freedom; those whose rights were of limited duration, had to hand over one-half. Even then, only the more prosperous peasants were affected, and since details had to be negotiated individually in each case, it was not difficult to spin out the process of commutation over many years. Meanwhile, a law of 1821 provided for the distribution of the commons as well as for the transfer of other lordly rights into a form of private property, again requiring peasant payments to the lords. By 1850, when the process had still not been completed, a further law was found necessary. Only now could poorer peasants purchase their freedom from services by making money payments, a process completed around 1865 (Mottek 1964, Wehler 1987).

The upshot was that the Junkers managed to add a great deal of peasant land to their own estates, while continuing to tie the rural population to the villages as cheap labour, and to preserve their extensive rights of jurisdiction. Some 420,000 hectares were gained by them in commutation payments, plus 300–500,000 hectares acquired in other ways — almost 2½ million acres in all. The typical agrarian form of organisation in the East therefore continued to be the Junker (landlord) estate, worked directly by the owner, using "free" wage-labour rather than rented out to tenant-farmers. Large units of this kind were well suited to adopt new methods, to supply overseas, above all British, markets while their owners were well placed to get legislation passed which favoured the large grain-producing estate specialising because of climatic and soil conditions largely in the production of rye.

Elsewhere in Germany conditions varied a great deal and are difficult to summarise. Typically, lords did not farm their land in the West, but let it out to tenants; in some regions the farmers were free-holders themselves. Commutation of services and dues therefore normally took the form of money payments and created an agrarian structure of smaller mixed farms, animal husbandry prevailing in the north-western regions of Germany. The areas occupied by the French were the first to abolish serfdom, Bavaria followed in 1808, Hesse in 1811. Against that, Saxony and Hanover made a start only after the Paris revolution of 1830. The abolition of the open-field economy and of the commons was delayed in Saxony, for example until 1832 and in most of Thuringia until after 1848, and this also delayed the introduction of modern methods. They were adopted all the readier thereafter as soon as private property rights in the land free of feudal obligations had been clearly established.

The regional differentiation of German agriculture is well brought out by the official statistics for 1907 showing the size distribution of holdings according to acreage. In East Elbia (including Silesia) 44% of the land was in holdings of over 100 ha (one hectare being almost 2½ acres) and a further 29% in holdings over 20 ha, whereas only 8% was in small holdings of 5 ha or less; by contrast, in the areas of partible inheritance, Baden, Württemberg and Grand Duchy of Hesse, 36% of land was in small holding of 5 ha or less and a further 45% in holdings of 5–20 ha, only 3% being in the very largest class. Much of the rest of Germany tended towards the middle, the mode being somewhere around the 20–50 ha range.

For a leading industrial country, the output, and especially the employment, in the agrarian sector remained astoundingly large (see Table VIII), productivity there being well below that achieved in the other sectors (Kuznets 1971b). This survival was owed in part to the privileges of the agrarians on the political plane and reflects the incomplete industrialization and modernization of the country to 1914. The improvements in agricultural productivity (Table IX) were achieved in a number of ways. The ending of the fallow year and the taking of new areas under the plough added some 40–45% to the acreage under tillage between 1820 and 1875. In those years the output of wheat and rye more than doubled, and the production of other grains also increased substantially. Thereafter cheap fodder grains

TABLE VIII
Structure of the German Economy 1850–1913[a]

	Shares in net domestic product, %			Shares in Employment, %		
	1850/4	1870/4	1910/13	1849/58	1878/9	1910/13
Agriculture	45.2	37.9	23.4	54.6	49.1	35.1
Industry, Mining	21.4	31.7	44.6	25.2	29.1	37.9
Transport	0.7	2.1	6.4	1.1	2.0	3.6
Services	30.0	25.0	20.5	19.1	19.8	23.4
Dwellings	2.9	3.3	5.1			
Totals	9.6[b]	15.7[b]	45.6[b]	15.1[c]	19.4[c]	30.2[c]

[a] Source: Hoffmann (1965), pp. 33, 35.
[b] Milliard Marks of 1913.
[c] Employment in millions.

TABLE IX
Annual Growth Rates in the German Economy 1850–1913, in %[a]

	Net domestic product	Numbers employed	Labour productivity
Agriculture	1.6	0.4	1.2
Mining	5.1	3.2	2.0
Industry	3.8	1.9	2.0
Commerce, Finance, Catering	3.2	2.6	0.6
Transport	6.4	3.3	2.9
Domestic Service	0.2	0.2	0.0
Other services, excl. Defence	2.0	1.8	0.0

[a] Source: Hoffmann (1965) p. 37.

were increasingly imported, and by 1910 Germany had to import 40% of her wheat. The number of animals increased spectacularly, with the exception of sheep, whose numbers declined. While in the first half of the century, output was expanded largely by adding acres, subsequently yields per acre were raised: between 1848/52 and 1908/12, the yields of the four major grains rose by around 60–70%. Artificial fertilizer, pioneered in Germany, had helped towards this achievement, and by the end of the period the most advanced agricultural regions had reached the efficiency levels of the then leading agricultural nations Denmark, Holland and England (Pounds 1985, p. 525). These were in the west; yields on the eastern estates with their extensive form of cultivation were well below them.

Technical progress required capital. Many Junker estates ran into heavy debt, a large proportion of them changed hands, and as they fell into the possession of non-noble urban families, the capitalistic profit-maximising tendencies on the land were strengthened. Medium-sized and smaller farms were helped by the mutual credit association of the Raiffeisen banks, developed after 1862, and the Schulze-Delitzsch co-operative and credit associations.

Much remained traditional also in the sphere of industry. The urban gilds, in particular, proved extraordinarily tough. "Gewerbefreiheit", the freedom to choose one's occupation, was introduced in 1808 in some parts of Prussia and in 1810–11 in the rest as part of the general reform movement, while in the regions occupied by the French the power of the old corporations had been completely swept away. However, in Prussia and other states, such as Saxony or Bavaria, the corporations remained on a quasi-voluntary basis, while entry into many occupations was in practice limited by licensing, by passing tests or even by requiring the agreement of the existing master craftsmen. The authorities welcomed the retention of some control as one means of keeping the unruly journeymen, forerunners of the later trade unionists, in check. Real freedom came only in 1868–9 to Prussia and the North German Federation, later incorporated in the Reich, but the craft gilds were by no means abolished even then. An Act of 1897 re-introduced a special registration category for artisan shops and the power to form "Chambers of Handicraft" with some compulsory powers if a majority of local masters demanded it.

It was, in fact, an aspect of the tenacity of the old, and of the survival of a peculiar German form of a "dual economy", that artisans

("Handwerk") kept something of an independent status and did not merge with the skilled working class, as in most other countries. For 1800, it was estimated that there existed 820,000 master artisans and 410,000 journeymen and apprentices, or half a dependent for each independent master; by 1870, the relationship had changed markedly: 1,070,000 masters faced 930,000 journeymen and apprentices, almost 1:1. Artisans, concentrated mainly in the food, clothing and building trades, kept roughly in step with the growth of the population as a whole (Table X). Though associated mostly with small workshops, handicraftsmen were found in increasing numbers also in large-scale industrial firms where it was more difficult to continue to think of themselves as members of the middle classes, which remained their recognised status.

The rise, from about the middle of the century onwards, of a modern industrial structure, marked by the early appearance of some giant firms, out of the mass of small-town, small workshop industry typical for Germany up to then, occurred with remarkable rapidity. While the speed was exceptional, the growth path conformed fairly closely to the standard model of such countries as Great Britain and Belgium.

In the early stages, it was the textiles which were the largest employers of industrial labour. Linen, a major traditional industry, failed to modernise in time and was largely wiped out by British competition, though some important centres remained. Wool, also, was slow to copy modern methods, but did so ultimately to survive. It

TABLE X
Structural Change in the Secondary Sector in Germany, 1800–1913[a]

| | Percentages of total employment | | | | Absolute |
	Putting-out	Large-scale industry Mining	Artisan workshops	All secondary industry	numbers, large-scale industry and Mining[b]
1800	9.0	1.5	10.5	21.0	0.12
1850	10.0	4.0	12.0	26.0	0.60
1873	6.0	10.0	14.0	30.0	1.80
1900	2.0	22.0	13.0	37.0	5.70
1913	2.0	23.0	13.0	38.0	7.20

[a] Source: Henning (1984) p. 130.
[b] In millions.

was the cotton industry and the much smaller silk industry in which mechanised methods caught on most successfully, even if belatedly. Hand weaving, using in part imported yarn, was organised on a putting out system until the 1880's. Thereafter, a rapid transformation created large, integrated firms.

The big spurt came with the building of the railways, followed by the discovery of the vast mineral reserves of the Ruhr district, in the 1840's and 1850's. The main carriers of this growth were the iron industry, coal mining and engineering. The output of coal, for example, rose from 1.3 million tons in 1820 to 5.1 million tons in 1850 and to 190 million tons by 1913, plus 87 million tons of brown coal, representing together about two thirds of the continental output. Pig iron output passed the 100,000 ton mark in 1826, the million ton mark in 1867 and by 1913 reached almost 13 million tons; steel rose from 126,000 tons in 1870 to 17.6 million tons in 1913. These figures were about double the British values. Production units were large, modern and efficient. By 1900, for example, no fewer than 24 Ruhr coal mines had an output of over 500,000 t each. In 1902, the average output per steel works was 75,000 tones compared with 40,000 tons in Britain.

Some of this expansion had taken place in the old iron and coal districts of Upper Silesia and the Saarland, but the new power house of German and European industrial might was the Ruhr region. Thus Ruhr coal output rose from 1 million tons in 1840–4 to 88.3 m t. in 1910–4. It produced 58% of all the coal mined in Western Europe (excluding the UK) and 71% of the coke (Pounds 1985). This region also saw the main concentration of engineering works, though the beginnings of German engineering had been scattered in accordance with the needs of the railways, and Berlin as well as Saxony remained important centres. Much of the textile industry and the heavy chemicals industry was also to be found in the Ruhr region.

By the end of the century, Germany had become the leading producer of certain types of chemicals such as dyestuffs and pharmaceutical products, reaching 85–90% of world output in certain branches. Electrical engineering was another strong point, two huge combines, Siemens and the AEG (the latter working largely with American patents) sharing much of the market at home and abroad. Altogether, Germany conformed well to the Hoffmann (1958) thesis that industrialization first affects the consumer goods industries, and turns to the capital goods industries in a second phase. Between 1850 and 1913, the annual growth rate of output in hard coal mining was 5.3%, in steel-making 6.6% and in metal working 5.6% as against the growth in textiles of only 3.0%, in clothing 2.6% and in the food industries 2.7%.[28] As far as labour productivity was concerned, it rose by +270% in metals, +67% in mining, +115% in textiles and +30% in the food industries.[29]

It may have helped the speed and ease of expansion of the Ruhr district that it was a "new" region, with few old towns, and those rapidly swamped by the immigrant population, to inhibit the expansionary drive of the new entrepreneurs. When urban authorities were founded, they were dominated by the industrial and mining employers and reacted to their needs. They put few obstacles in the way of the tough, ruthless, single-minded entrepreneurs of the German industrialization and "foundation" years.

Neither were employers in the growth industries hampered by labour shortages. Up to about 1880, immigrants came largely from nearby rural areas. After it, huge waves of immigrants, including Polish-speaking ones, came from the impoverished East Elbian pronvinces[30], diverted from their former target areas in the USA by the employment opportunities in the heavy industries of the Rhineland and Westphalia. Something of the same kind occurred in the working-class suburbs of Berlin, which grew fastest of all. By contrast, growth was slower in the older industrial districts of Saxony, Upper Silesia or the Wupper valley (Tipton 1976, Hoffmann 1963). It is possible that the rapidity of the change itself contributed to the survivals and the "dual economy" structure of Germany[31] in the decades to 1914.

The Railways as Leading Sector

Railways were one sector in which Germany was ahead even of France (Table V). This is all the more surprising in view of the lack of political unity, so that the early lines radiated from numerous minor centres instead of being planned, as in France, on a national basis.

Apart from a few strategic lines in Prussia, the German railways essentially followed economic logic and conformed rapidly to national needs as sketched by such dreamers as Friedrich List early on. In a few cases the State built the early lines, in others they were built by private companies, and it may be that this mixed initiative contributed to their speed of development.

Building started in earnest in the 1840's after a few pioneer lines had

been completed in the 1830's, and continued at a high level for the next
three decades. Its importance for the German industrial revolution is
hard to over-emphasise: "German industrialization is seen . . . as a
process of unbalanced growth with the railways as the leading
sector".[32] As a proportion of the nation's net investment, they reached
11.9% in 1951–4, 19.7% in 1855–9, and remained at that level until
peaking at 25.9% in 1875–9 (Fremdling 1985). At first, the needs of the
railways for iron and machinery greatly exceeded Germany's productive capacity, and much of these were imported from Britain and
Belgium. But within two decades, railway demand, together with
developments elsewhere, had created, first rail rolling capacity, and by
the 1850's, also iron making and locomotive building capacity, which
laid the foundations of later growth. In the 1850's, railway demand for
iron constituted one-third of domestic production, and one-quarter of
domestic consumption of iron. By 1860–5, the position of Germany as
an importer of rails had been changed: over 23,000 tons of rails could
be exported; by 1866–71, this had reached 150,000 tons (Fremdling
1977, 1985, Henderson 1975).

At no time does there seem to have been a shortage of capital for the
railways, though the banks, unlike Britain but similarly to France, had
to take much of the initiative. Interest guarantees by the State, as in
France, played a part in the building of the less profitable lines, and by
1914, almost the whole of the network had been acquired by the
State(s), its annual surplus furnishing around 10% of their revenue. As
elsewhere, railway shares were significant in the development of share
holding and speculation. An indication of the importance of the railways as carriers, especially for regions away from rivers and coastline,
may be given by the fact that ton-kilometres of freight carried rose
from 1.7 milliard in 1860 to 13 milliard in 1880 and 61,7 milliard in
1913; passenger km in milliard rose from 5.7 in 1873 to 41.4 in 1913
(Henning 1984, Fremdling 1985).

Banking and Finance

As was appropriate to a traditional economy, the German banking
system to the mid-century was rather primitive. Private bankers
conducted business for the numerous princes, or provided short-term
finance and exchange services for traders. There were savings banks for
the lower middle classess, supported by the authorities, and the
Prussian "Landschaften" to provide loans on mortgage to landlords.

Early examples of central note-issuing banks were the Bavarian Mortgage and Discount Bank (1835) and the Bank of Leipzig (1838), but the first important step towards a more modern approach was taken with the transformation of the Royal Bank in Berlin into the Bank of Prussia, with rights of note issue, in 1846. On the formation of the Empire it became the Reichsbank, a central bank of the modern type with the task of maintaining the newly established gold standard. Unlike the Bank of England, it had numerous branches, and it stood ready to help other banks in times of stringency, which in turn permitted to remainder of the banking system to be more adventurous in its lending policy.

Possibly the most widely debated aspect of the German banking system was the rise of the so-called "universal" or "mixed banks". An early example was formed by the conversion of the Schaaffhausener Bankverein in 1848. It was followed in the 1850's and 1860's, in imitation of the French *Crédit mobilier* and against the initial hostility of the Prussian bureaucracy, by other joint-stock banks. These included the Darmstädter Bank (1853), the Discount Bank (1856) and the Berliner Handelsgesellschaft (1856). Among the founders of each were established private bankers. These banks, working with their own capital rather than with customers' deposits, were willing to make longer-term investments in industrial firms, to undertake their share issues, carrying some of the risks themselves, and therefore to take a more direct management interest in industrial firms than was common in Britain or even in France.

Such services were of particular value in rapidly expanding industries, especially those requiring much capital, as in iron and steel, coal mining, engineering and chemicals. According to the Gerschenkron thesis, they compensated for the slower growth and richer capital base available in Britain[33], and were, according to others, instrumental in sustaining the rapid rate of German industralization. The banks were credit-creating and by 1914 also had an expanding deposit business. Others noted that smaller firms, and industries like textiles, were neglected by the banking system which therefore may have contributed to a distortion of economic growth (Gerschenkron 1966b, Eistert 1970, Tilly 1967, Neuburger and Stokes 1974, Fremdling and Tilly 1976).

The paid-up capital of German joint-stock companies rose from 548 million Marks in 1857 to 3,339 million Marks in 1900 and 11,424 million Marks by 1913.[34] Banks themselves amalgamated, and by 1913 the scene was dominated by the "Big Four". The early rise of the very large company elsewhere, often vertically integrated and controlling much social capital such as housing, was connected to the backing it received from its associated bank or banks. Moreover, as each bank acquired interests in a number of firms in the same sector, it made its influence felt to induce its clients firm to collaborate with others in its branch of trade and cartellise sales and possibly production.

Germany was the land of cartels, helped not least by the fact that Governments supported them, unlike the Governments of Anglo-Saxon countries. A legal decision of 1897 strengthened them further, and in 1910 the potash cartel was even made compulsory. Considered to be "children of the depression" of the 1870's, some 385 cartels were counted in 1905, and by 1907 they controlled 50% of crude steel output, 74% of mining and 90% of the paper market.[35] Aided by tariff protection, they could in slumps maintain their output by dumping, but it is not clear that his helped the German economy as a whole. Sometimes foreign producers of finished goods, like British ship-builders, received their semi-manufacturers more cheaply from Germany than the competing German shipyards. German firms were by 1913 involved in over 100 international cartels and shipping conferences. Theory might maintain that competition was good for efficiency, but the cartellised industries were among the most successful in terms of growth and international competitiveness.

The Export Economy

There are no satisfactory statistics relating the German foreign trade in the early years of the nineteenth century, nor is it clear whether, or to what extent, trade among the sovereign states which later formed the German Empire should be considered foreign trade. With the formation of the *Zollverein* on the first day of 1834 the statistical base improved, at any rate for the increasing parts of Germany embraced by it, and the main outlines become clearer. In the early years of German industrialization, "Germany" imported largely tropical products, raw materials and semi-manufactures, and exported foodstuffs and finished manufactures, whose share rose to 55% of exports in the 1850's.

It showed at that time in classical form the trading pattern of an industrializing country sitting between advanced economies on one side, and more backward economies on the other. Germany made use of her strong connection with Britain, as well as Belgium and, to a lesser extent, France, to obtain machinery and locomotives, as well as semi-manufactures, like machine-spun yarns and pig iron, produced cheaply by the most advanced methods then available. The finished goods produced with their aid such as textiles and iron, she could sell to the countries to her eastward, and even to westward and the USA, since her skills were still combined with low wages. An important item of sale was grain as well as naval stores from the eastern provinces to Britain, which became part of a triangular trade, the other two legs being British semis to Westerns Germany, and manufactures thence to East Elbia (Pollard 1981, Fremdling 1977, Tilly 1978).

The *Zollverein* had, on the whole, kept to low to moderate tariffs, and in the general lowering of European tariff barriers introduced by the Anglo-French commercial treaty of 1860, it joined the network of agreements with most-favoured-nations clauses. By the 1870's, Germany had become almost a free-trade country, but this trend was reversed with the tariff law of 1879, which was soon followed by ever higher protectionist barriers.

The causes of the 1879 law were partly to be sought in party political manoeuvrings, and were partly fiscal, since the income from external duties was one of the few sources of revenue available to the Reich as distinct of the state governments within it. But it had also a solid economic foundation in the depression of the 1870's which shattered German hopes of uninterrupted industrial progress, as well as beginning to flood Europe with cheap overseas grain. Protecting industrial products as well as grain, the law of 1879 has been described as a compromise between rye and iron, but the iron and steel industry which was rapidly becoming the most efficient in Europe, had in fact little need of it. The real drive behind it came from the class of large landowners, who used their political power to keep out foreing grain in order to prop up their own selling prices, even though they were able to supply the German population with an ever smaller share of their preferred wheat grains.

German manufacturers thus had to remain competitive in world markets while the cost of living was kept up artificially by the tariff. Though compensated, in some cases, by duties on their own products also, the industrialists and their political representatives, as well as the representatives of labour were very conscious of this handicap and

sought to reduce it in a powerful campaign from about 1890 onward.
In defence of the *status quo* the agrarians, cleverly allying themselves
with the smaller farmers who had, in fact, little to gain from protection
which raised the prices of their fodder grains, developed the notion of
the "agrarian state" versus the "industrial state" (Barkin 1970,
Lebovics 1967). A large agricultural sector, they maintained, should
survive for social as well as military-strategic reasons, even though on
strict economic criteria it would have paid Germany to import more of
her food and have paid for it by industrial exports. Their political
power ensured that a large, antiquated and in part inefficient agricultural sector was preserved until well after 1914.

The share of foreign trade in the German economy was considerably
less than in the smaller European countries or in Great Britain,
amounting to 9% of NDP in 1835, 18% in 1873 and remaining at or
below that level until 1914. But, because of the weight and size of the
German economy, and the concentration of German exports on certain
manufactured products such as machinery, metal goods, chemicals
and machine-produced textiles, they came to dominate European trade
and become a major factor in world exports, while the country formed
a leading market for such imports as iron ore from Sweden or agrarian
products from Eastern Europe. Germany's share in world trade rose
from 9.2% in 1860 to 11.0% in 1891-5 and to 12.3% in 1913 (Kuznets
1966); her share of the world's manufactured exports from 19.3% in
1880 to 26.4% in 1913 (Tyszynski 1951).

Germany used little foreign capital in her own industrialization. Her
capital exports began late but grew with considerable vigour in the
years to 1914. The banks which were active in this field were subject to
some political direction, while political objectives cut down the lending
to Russia in the 1880's, and to Serbia in 1893-1906 (Trebilcock 1981,
Feis 1964). Total foreign investments have been estimated at 7.2
milliard Marks in 1882, rising to 20 milliard Marks in 1913.[36] In the
early years of the twentieth century, when half the British capital
formation went abroad, Germany invested only about 6% of her
capital formation in that way (Landes 1969).

The Role of the State
The traditional picture of Germany is one of an interfering state and an
active bureaucracy. Much of the literature emphasises the positive role
played by German Governments, especially that of Prussia, in the early

stages of the industrialization process. Beside the buidling of roads and canals, the import of British technology to State coal mines and metal works, the visits to Britain and invitation to foreign technicians, direct initiatives are also mentioned, like the industrial investment policies of the Prussian " Seehandlung" which acted, under the guidance of its officials, as a kind of enterprise board in the first half of the nineteenth century (Henderson 1975, Fischer 1963). The more recent literature is less positive. (Trebilcock 1981, Borchardt 1973). While it is true that the State built some of the earliest railways in Hanover, Baden and some other states, the Prussian bureaucracy was suspicious and obstructive, as it was of early joint-stock banking (Henderson 1975). The freeing of access to most occupations was anything but smooth, as we have seen, and State control over mining in Prussia, which had some positive features as well as inhibiting expansion, was brought to an end only in 1851-65. Whatever the ultimate benefits of the *Zollverein*, it was certainly not intended as an aid to industrialization, though its tariff policy came in course of time to benefit some of the key industries, like the basic and secondary iron industry, freed raw materials while taxing manufactures, and it did have the clout to get better terms for German interests in international trade negotiations.

The fact was that even the more advanced German countries like Prussia were dominated by an agrarian class which was only too aware that industrial progress and its associated rise of the urban middle classes would threaten its hegemony. Civil servants sent from Berlin to administer the industrialized Western provinces which had fallen to Prussia after 1815, showed little sympathy for the special needs of their charges (Tipton 1976).

Above all, as we have seen, the agrarian ruling class was able to burden the economy with a grain tariff even after Germany had turned into a highly successful industrial exporter — in significant distinction from the British experience of 1846. Tariff rates for wheat and rye, fixed at 10 Marks per ton in 1880, rose to 50 Marks in 1887-91 and after a temporary reduction, to 70-75 Marks in 1906-14. These rates were reflected exactly in the excess of grain prices in German cities over world prices (Clapham 1963, p. 211). Since the German population was increasingly unwilling to consume the home-grown rye, the Junkers managed by a complex export bonus system (1894 and 1902) to have their exports of it subsidised. They prevented the building of an

East-West canal, and taxed grain traders as "usurers", while creating special banks for themselves in 1895.

A special word must be said about education. The German states were among the first in Europe to introduced compulsory education, though their reasons were the creation of an obedient, god-fearing rather than an economically efficient population. Aided by compulsory military service, German workers were found to be well disciplined and well prepared to acquire industrial skills when the time came. Morever, the German university ideal with its emphasis on research turned out to be an excellent aid to "high-tech" industries, such as chemicals and electrical engineering in the later nineteenth century (Landes 1969).

Against this, the accounts in the literatures of deliberate State subsidies, of diplomatic and colonial drives to aid German industry, are greatly exaggerated. "The authorities were inclined rather to tolerate the evolution of industry . . . Frequently so-called policies of industrialization turned out to amount to no more than the belated creation of conditions which had existed in other western countries long before their industrialization".[37] It was precisely the strength of the German industrialization process that it depended on market success, rather than on pampering by the State.

2.2. The USA: Industrialization on a Primary Commodity Base

The American Model of Industrialization

Unlike the countries discussed so far, but like Russia after her, the United States entered upon the stage of industrialization long before her resources of land and space were fully used up. Her industrialization phase was marked, therefore, by a simultaneous lateral expansion into empty territory while she maintained, secondly, a large export surplus of primary commodities, in which she enjoyed a relative advantage on an international basis. But thirdly, unlike Russia, the USA started from a very high level of income per head and of economic sophistication, based on her rich resource base. These, then, were the characteristics of the American model: a high starting level, industrialization rooted, in the early phases, on the primary sector, and extensive widening simultaneously with deepening of capital.

As early as 1800, a standardised measure of per capita industrialization showed the USA roughly at the same level as France, Belgium and Switzerland. By 1860 she was ahead of France, though behind the other two; by then she supplied 7.2% of world industrial output. In per capita raw cotton consumption, she was only second, behind Britain, in coal consumption only behind Britain and Belgium, and in fixed steam power, as well of course as in agricultural production, she already led the world[38] — though she had, at most, only just "taken off" on her industrial revolution. The American growth of productivity was no faster than that of other major economies engaged in the process of catching up (Abramovitz and David 1973), though it was faster than that of Britain; but given the high starting level and the simultaneous rapid expansion of the population boosted by an exceptionally high level of immigration, the United States was bound to emerge, at the end of her comparatively short transformation, not only the richest, but also the largest industrial power of the world. By 1913 she accounted for 32% of the world's manufacturing capacity (Bairoch 1982). It was clearly the success story of the nineteenth century.

In 1800, the proportion of the labour force in agriculture exceeded 80%; by 1840 it was still 68%, in 1860, 53%, and only in 1890 did it drop to 43% and to 32% in 1910 (Kuznets 1966, Lebergott 1984). In terms of commodity output, agriculture accounted for 72% in 1839, 56% in 1859 and 37% still in 1899. To this has to be added the produce of mining, 1% in the earlier years and 4% in 1899.[39] It is clear that high agricultural productivity played a major part in the high level of American incomes, that much of manufacturing consisted of processing agrarian products, and that the transfer of resources out of agriculture into manufacturing after 1840 had its share in boosting overall growth (Temin 1975, Gallman 1966, North 1966, Lebergott 1966). Industrialization mainly on a primary product base was the model also followed, *mutatis mutandis*, by such countries as Canada, Australia and the Scandinavian countries.

Associated with it in the USA was expansion westward, thus raising agrarian output without the increased marginal costs found elsewhere. Industrialization took place "against a background of territorial and agricultural expansion, interwoven with price and other impulses from the international economy."[40]

Rostow initially placed America's "take-off" in the period 1843-60, that is, rather later than continental north-western Europe, though for New England alone it was 1815-50 (Rostow 1960, 1978). It is not, of course, impossible to treat the USA as a continent rather than as a country. North agreed that 1820–60 was the "era of industrialization", and that by 1860 "the problems of industrialization were behind, and the further development of manufacturing was not so much a question of how it could develop, but rather its effects on the American economy and society". Hacker also believed that the USA was an industrialized nation by 1860, and Jones held that "the United States breakthrough had already occurred in the 1840's and 1850's".[41] Against this, Chandler thought that the "significant transformation" of the American economy had to be dated from the depression of the 1870's (Chandler 1959) and in later writing Rostow himself opted for 1843–70 (Rostow 1978). The level of capital formation, which jumped in the 1850's (Table XI, Abramovitz and David 1973) seems to lend support to the Rostow view, though the level of investment was raised again after the Civil War. Others failed to see any special acceleration deserving the name of take-off in the 1843–60 period (North 1966, Fogel 1970, Fogel and Engerman 1971).

The Civil War interrupted all existing trends. At one time it was seen as a positive force, encouraging innovating entrepreneurship, particularly in manufacturing. Recently, its destructiveness has received more emphasis (Hacker 1970, Goldin and Lewis 1975, Temin 1976). If the catching-up process, back to the original growth trend, is thought to have been responsible in part for the observed acceleration, then that due to the "take-off" could not be separated out.

Yet if America was "backward" in the sense that her "take-off" was a generation behind the British, her "backwardness" was of a different kind from that of the European continentals. True, like them, "the US

TABLE XI
Gross Capital Formation in the USA, 1834–1908[a]

	GCF as % of GNP	GCF in $ bn of 1860	Improved farm land, $ mn of 1860	GCF in $ bn of 1860	
1834–43	9	.141	133	1839	.200
1849–58	14	.474	118	1849	.258
1869–78	22	1.37	147	1859	.532
1879–88	22	2.36	102		
1889–98	28	4.04	96		
1899–1908	28	6.17	89		

[a] Source: Gallman (1966) pp. 11, 34, 35.

in the nineteenth century . . . was a major beneficiary of the technological progress which had already taken place in Great Britain. In large measure, her economic development in this period involved the transfer and exploitation of British techniques".[42] Moreover, the manufacturing sector might have employed only 2.8% of the working population in 1800, 8.8% in 1840 and 13.8% in 1860 (Lebergott 1966) and in that sense the country was less industrialised than Britain, the leading country, in 1820–60. But such manufacturing as there was, was fully up-to-date given the needs of the economy. Technical innovations from Britain were absorbed quickly and, unlike the continent, were not held up by craft or technical incompetence. In some significant fields, the Americans not only drew level but pulled ahead.

A good example was the building of steam engines. It was the Americans who pioneered the application of high-pressure steam and the steam propulsion of river boats. By 1838 there were over 100,000 steam HP in use, of which 57% were in steam boats and 7% on railways, and some 250 markers of steam engines, scattered over all regions. By 1860 the western rivers carried a steam tonnage of 162,000 t and the USA led the world in per capita stationary steam horsepower (Temin 1966, Rosenberg 1972, 1976, Bairoch 1965). In railway building, British technology was accepted almost at once.

The cotton industry, much the largest manufacturing sector, had become an industry of large units centred on New England: output increased from 46 million yards in 1806 to 323 million yards in 1840 and 857 million in 1860 (Zevin 1971, North 1966). Most interesting, perhaps, was the rise of the "American system of manufacture", the evolution of labour-saving mass production techniques with interchangeable parts, which had begun in armaments manufacture in the first decade of the century and had reached a level by the 1850's which British observers considered to be superior to their own (Hounshell 1984, Rosenberg 1969, 1972, 1976). In timber shipbuilding the Americans also led the world (Jones 1965). If coke iron production lagged, and much rail iron had to be imported from Britain in the 1840's and 1850's, there were good economic grounds for the American preference for charcoal (Temin 1964).

Whether high American wage costs, caused in turn by the easy availability of land, was responsible for the peculiar and early American predilection for labour-saving machinery, has been a matter of some controversy since the thesis was first comprehensively put forward by

Habakkuk (Habakkuk 1962, Hacker 1970, pp. 149–50, Temin 1975, Abramovitz and David 1973). American wages were indeed higher (Rosenberg 1976, chapter 3, Bairoch 1965, p. 1096, Brown and Browne 1968), but so, most likely, was American labour productivity (North 1966). US machines tended to be flimsier, hence they were renewed more frequently and were more up-to-date; they also used more wood than the British, as it was relatively cheaper. But it is not clear why they should have substituted capital for labour, since capital, though plentiful, (Table XI), was also dearer than in Britain.

Given that the per capita income level was no lower and may well have been higher than the British, if food consumption and its share in household expenditure are any guide, it is not clear that the USA economy, though differently structured, was noticeably behind the British even before 1860. After 1865 it quickly caught up and then overtook Britain, even by conventional measurement.

The Transport Revolution
The breakthrough of the American economy into a modern form has been associated in the traditional literature with the building of the railway network. In the Rostow scheme, for example, the railways were the most important "leading sector" in the period 1843–60. This view has been strongly challenged (Fogel 1970) but has survived nonetheless. In American conditions, the railways were not only involved with the industrialization process by forward and backward linkages as, for example, in Germany, but also with the simultaneous expansion westward, the filling-up of the empty spaces which was such a vital feature of the American model of industrialization.

Before the railways, a vigorous period of canal building, beginning in 1803, had helped to bridge the gap between the areas east and west of the Appalachians which had hitherto had their separate system of communications, the coastal waters, and the Ohio-Mississippi systems respectively. The Erie canal, completed 1825, provided the first major link and, incidentally, ensured the rapid early growth of New York city. By 1860, 4,250 miles of canals had been built at a cost of $190 million (Ransom 1964). The peak investment rate was reached in 1838–41[43], by which time, however, the railways had begun to reduce canals to secondary importance.

Railway building may be conceived of as having occurred in four overlapping waves: In the first decade, there were mainly short lines in fairly densely settled regions. In the years to 1860 followed a second phase, in which a denser network was established in the East, as well as to some extent in the middle west, with some longer lines in the South, with links between the three regions. A third phase, beginning in the 1860's, led to the completion of transcontinental lines to the Pacific coast, across largely empty space. A fourth bout, with a peak in 1906, then filled in the remaining gaps.

In the first building phase in the 1830's 280 miles a year were completed, rising to 368 miles in 1840-7, and 1824 miles a year in 1848-60. Building then slowed down, to average a mere 815 miles per annum in the war years, but picked up in 1866-73, to an almost incredible average of 4,436 miles a year, a peak of 7,439 miles being reached in 1873. There was some slowing down after the crash of that year, but still an average of 3,312 miles was added every year to 1880. A general overview may be gained from Table XII.

The stupendous effort represented by a concentrated capital investment of this nature may perhaps best be put in perspective by noting that in 1840, the USA had built about 60% of the world's mileage; it was still 46% in 1860 and 44% in 1888.[44] This was carried through, in part, by building the lines flimsily on first opening, by attracting some foreign capital for them, as well as by generous public

TABLE XII
Railway Building and Investment in the USA 1830–1890[a]

	Miles of railway operated at year-end	Property investment in railways, $ mn
1830	23	
1835	1,098	
1840	2,818	
1845	4,633	
1850	9,021	318
1855	18,374	764
1860	30,626	1,149
1865	35,085	
1870	52,922	2,477
1875	74,096	4,658
1880	93,262	5,402
1885	128,320	7,843
1890	166,703	10,123

[a] Source: US Bureau of Census (1960).

aid. Nevertheless, the question has been asked whether the widespread fraud, chicanery and diversion of public property into private pockets associated with the building of the major American railway lines had not led to a misapplication of resources which would have yielded more if they had been applied in the development of the settled area, leaving the transcontinentals until later. The majority opinion considers the money well spent, in bringing in immigrants, furthering agriculture and mining, and creating competition in many fields earlier than would otherwise have been the case (Lebergott 1984). In particular, the inter-regional links, allowing more specialization in the North-East, the South and the North-West before 1860, should be stressed here.

Certainly, without Government aid the system would have taken much longer to build up. Individual states, which were eager to divert the main lines in their direction, once building had been determined on, provided land grants, possibly the most effective aid, since the land gained in value as soon as a line was built. There were also monopoly rights, tax exemptions, free rights of way and permission to open banks by the states, among other favours. The federal government helped by land grants as well as by surveying the land and allowing tariff rebates on imported rails. All told, the railways were calculated to have received $1.5 billion in aid from the public authorities, though these also received some benefits in the form of dividends repaid and rises in land values (Hacker 1970, Jones 1965).

In the high boom years of the 1870's, some 20% of American investment went into railways, and in the 1880's it was still 15%. The railways also had their share in fostering the iron industry, though up to the Civil War much of the railway iron used came from abroad: only by 1858 did the quantity of home produced rails exceed the tonnage imported. The share of domestic pig iron consumed by the railways rose from 9.1% in the 1840's to 22.4% in the 1850's and a maximum of 25.4% in the years 1855–60.[45]

One further impact of the railways on American industrialization remains to be considered: that was as the training and experimentation ground for large-scale management, for the organisation and combination of large masses of capital, and for acquainting large sections of the population with shareholding and the speculative possibilities of the stock exchange (Chandler 1977).

Agriculture and Mining

In the primary sector, it was raw cotton which acted as the main export provider and currency earner before 1860; after the Civil War, wheat and minerals were added. It has even been possible to hold that the speed and tempo of the country's economic development to 1860 depended in the end on the stimulus of the cotton exports from the Southern states, which rose from 83 million lbs in 1815 to 198 mn lbs in 1830 and 1768 mn lbs in 1860. In the decades from 1830 to 1860 cotton formed, year by year, between 50%–60% of total US exports by value (North 1961, statistics on p. 233 and Rostow 1978, p. 132). Meanwhile the United States share in the world's cotton production rose from 16.3% in 1811 to 66.0% in 1860.[46]

Cotton was a product largely of slave labour. Although the South was also a major producer of other commodities, accounting for most of the tobacco, around half the corn and the hogs, and considerable proportions even of the wheat and dairy cattle of the United States, it was thought that its prosperity depended on cotton and that, in turn, depended on slavery on the one hand, and the power to expand by incorporating new lands in the West into the slave system, on the other. Certainly the richer white slaveowners, though only a tiny minority of the population, dominated Southern politics.

The issue of slavery in connection with American economic development has been much debated. According to the traditional view, slavery obstructed development since it encouraged maximising acres and slaves instead of profits and capital accumulation, it held back the installation of modern equipment and it wasted manpower either in unmotivated work on the part of slaves, or in the need for large numbers of slavedriving supervisors. It could certainly be shown that in terms of manufacturing capacity[47] and railway mileage, the South was far behind the North, and planter society was heavily in debt to the North and to Britain.

Against this, some modern research had emphasised that slavery was essentially profitable right up to 1860, that slave owners thus acted rationally, that output per head was not inferior, and might even have been superior to that in northern agriculture and that it had proved possible, even within slave society itself, to introduce systems of motivation more subtle and effective than the time-honoured brute force (Fogel and Engerman 1974, "Slavery" 1967). Without the Civil

War, it could thus be argued, slavery might have been viable for a long time yet.

There is no disagreement about the drastic reduction of incomes and the depressed condition of the South after 1865, though by 1880 the quantity of American cotton exports exceeded the record sales of 1860 and stayed above them thereafter. Three main reasons have been made responsible for this overall decline: the actual destruction caused by the war itself; the rise of alternative cotton supplies abroad, in part called forth by the war-time "cotton famine"; and the failure, for many years, to achieve the productivity under free labour which had been achieved by an integrated plantation economy (Ransom and Sutch 1977, Wright 1974). Whereas the per capita income index of the South, with US = 100, was 76 in 1840 and 72 in 1860, it stood at only 51 even by 1900 (Easterlin 1960).

Northern agriculture expanded both by opening up new territories in the west, as well as by improvement in methods and technical equipment. The occupation of former forest land, followed further west by prairie and plain and ultimately by near-desert land required capital investment, but in the end productivity was no lower, and at times even somewhat higher, than in the original territory of the New England and Middle Atlantic states, so that expansion, unlike the experience of European countries, could proceed without diminishing returns. A yearly average of 11.4 million acres was put into farms in the 1850's, 12.8 million acres in the 1870's and 8.7 million acres in the 1880's. The number of farms increased in this period from 1.45 million to 4.56 million, and to a further 6.45 million by 1914 (US Bureau 1960).

The western lands needed innovations like the steel plough, barbed wire and water pumps to make them cultivable, and throughout, there was substantial improvement in productivity per man by such means as increased use of fertilizer and farm machinery, of which the mechanical reaper, adopted widely in the 1850's, was only the most spectacular (David 1971). Value added per gainful worker in agriculture róse from $212 in 1839 to $251 in 1859 and $358 in 1899; conversely; the man-hours required per 100 bushels of wheat fell from 373 in 1800 to 152 in 1880, and for 100 bushels of corm (maize) from 344 to 180. Total labour productivity in wheat production in 1910 rose to 417 (1840 = 100), and of the increase of +317%, mechanisation was responsible for +146%; similarly, in corn production, of a labour productivity increase of +265%, mechanization accounted for +127%.[48]

The result of the multiplication of an enormous expansion of persons and acres engaged, with their gain in productivity, was a spectacular increase in total output. Farm gross output at constant prices rose from $343 million in 1800 to $4,129 million in 1880 and $6,409 million in 1900, or almost 19-fold in the century. In similar constant dollar values, the output of eggs rose 50-fold, of wheat 23-fold, of corn almost 15-fold and of "truck crops", or market gardening produce no less than 124-fold.[49] The rise in the American share of the world's wheat output from 18.8% in the 1870's to 27.9% in 1894–9[50] reflects both the rising importance of that item in US exports, and the devastating effect of it on European agriculture. From one point of view, the strength of the American economic dynamic is to be sought in this sector until well into the last third of the nineteenth century (Hughes 1970, chapter 4).

The United States was fortunate in her mineral resources. Her rich coal reserves, both bituminous and anthracite, were brought to light step by step as the population moved westward, and helped to make up for the increasingly costly wood supplies. Per capita consumption of all fuels thus remained steady, at 3.9 net tons coal equivalent in 1850 and 3.8 net tons in 1880, at a time of improved fuel economy. Bituminous coal output rose some thousand-fold from 108,000 tons in 1800 to 4 million tons in 1850 and 111 million tons in 1890; total mineral energy consumption, including oil from the late 1850's onward, rose from 219 trillion B.t.u.'s in 1850 to 4,474 in 1880 and 7,322 in 1900, or a 33-fold increase, while fuel wood consumption remained steady at just over 2,000 trillion B.t.u.'s (US Bureau 1960).

Among other mineral resources, vast supplies of high-grade iron ore were found along Lake Superior, which were even exceeded after 1900 by the astonishingly rich Mesabi range. Gold, silver and a wide variety of other minerals were also found. By 1909, the USA was much the largest copper producer in the world, as well as producing 30% of the world's lead, 34% of its zinc and 17% of its gold.[51] Development was never hampered by lack of mineral resources, but on the contrary frequently driven forward by lucky finds.

Economic Growth
From the middle of the century onward, manufacturing carried on

increasing share of the growth momentum. Table XIII shows that it
represented little more than 15% of commodity output even in 1839,
but its share was constantly on the increase, exceeding half the
commodity output by the end of the century. Though forming still a
relative limited sector of the economy compared with the industrial
European countries, in absolute size it began to form a major or
dominant segment of the world's manufacturing capacity in a growing
range of products.

Typical here was the rise of the modern iron and steel industry. At
the time of the Civil War, the USA had only just emancipated herself
from dependence on British iron imports for rails. The Bessemer steel-
making process, invented in the UK which revolutionized the industry,
was paralleled by William Kelly in America, and was first applied
commercially there in 1866; before long it had been improved by a
number of American patents, and exceeded the British in efficiency.
The open-hearth-process came relatively late to the USA, in the 1880's,
but its costs were reduced rapidly to the level of the older process.
Increases in size and a whole range of innovations followed. Pig iron
output rose from 130,000 tons in 1828 and 750,000 tons at the time of
the outbreak of the Civil War, to 4.6 million tons in 1882 and 23.6 mn t
in 1910; steel output from 20,000 tons in 1867 and 1.7 mn t in 1882 to
23.7 mn t in 1911 (Temin 1964). By the turn of the century, the USA
had become much the largest and most efficient producer of this key
material.

Cotton manufacture was another industry in which American
innovations, such as the ring spindle and the Northrop loom, led the
world. Raw cotton consumption, running at 845,000 bales in 1859 had
more than doubled by 1881 and quadrupled to 1900, at 3.69 million
bales (US Bureau 1960). But it was in capital goods, especially in
machinery and locomotive production, that expansion was most
dramatic and significant, rising in current prices from 18–20% of
commodity output in 1839 to 31% in 1879 and 34–36% in 1899.[52]
Steam horsepower in factories amounted to 36,000 HP in 1838, 1.7
million HP in 1860 and 10.3 million HP in 1900, almost a 300-fold
increase (US Bureau 1960). In total industrial potential, the USA at an
index of 127.8 (UK = 100) had by 1900 overtaken the hitherto leading
power (Bairoch 1982, Brown and Browne 1968, p. 59). There is little
doubt also that in company organisation (Chandler 1977) and in the

TABLE XIII
American Product and National Income 1839–1899[a]

	G.N.P. $ billion Constant Prices of 1860	Commodity Output $ billion Constant Prices of 1879	Value added in Manufacture Constant Prices of 1879 $ billion	Rates of growth Gross Annual Product % per annum			Frickey's Index of Manufacturing
				Product	Product per capita	Product per worker	
(1800–1840)				4.29	1.27		
1839	1.62	1.1	190	4.24	1.10	0.64	
1849	2.43	1.7	488	4.95	1.80	1.71	16[b]
1859	4.10	2.7	859	1.99	−0.39	−0.08	25
1869	6.40[c]	3.3	1,078	4.95	2.56	1.88	36
1879	10.6[c]	5.3	1,962	3.73[d]	1.44[d]	1.00[d]	66
1889	14.4[c]	8.7	4,156	4.04[e]	2.20[e]	1.50[e]	100
1899	21.8[c]	11.8	6,262				

[a] Sources: Kuznets (1971) p. 18; Gallman (1960) pp. 16, 26, 1966, p. 26; Rostow (1978) p. 388; Williamson (1964) p. 272, North (1966) p. 673.
[b] 1960.
[c] Averages for the decade, starting with the given year.
[d] 1878/82–1888/92.
[e] 1888/92–1898/1902.

creation of large merges and trusts, the USA had by the 1890's taken the lead over Europe.

The increase in total GNP is shown in the first column of Table XIII. At constant prices, GDP increased annually at a rate of 4.4% in 1820–70, and 4.1% in 1870–1913, representing a per capita rate of 1.4% a year and 2.0% a year respectively, much faster than any other comparable country with the exception of Japan.[53] Part of this result was achieved by a considerable increase in capital. The productivity rise beyond the enlarged inputs of labour and capital, or total factor productivity (TFP), has been calculated to have increased by the respectable rate of 0.3% a year in 1800–55, rising to 0.5% for 1855–1905 and 1.5% for 1905–27. By a different calculation, TFP was found to have risen by 1.5% a year in 1889–99 and 1.1% in 1899–1909 (Abramovitz and David 1973, Kendrick 1961).

Some of the causes for the exceptionally rapid and successful completion of the phase of industrialization in the USA have already been named. They include rich resources in land and minerals and, in some key areas, good natural means of transport. Among other causes were a tradition of freedom unhampered by surviving feudal or landed interests, yet circumscribed by a rule of law favouring those in possession; a vigorous tradition of entrepreneurship encouraged by the high prizes, economic and political, beckoning to the successful business-man; and an exceptionally high level of education.

Education is difficult to measure, still more difficult to evaluate: much of the education provided in the schools of, say, traditional Prussia, would be of little help in the economic development of a nation. Certainly, the USA spent more on education than Europe and a higher proportion of its population was enrolled in schools (Fishlow 1971, Rosenberg 1972, p. 38): by 1860, many states had compulsory school attendance and, by 1890 school to the age of 14 was common, though it was found that even in 1870, the average school attendance was only 4 months in the year (Lebergott 1966). The Morrell Act of 1862 establishing Land Grant Colleges fostered the training of engineers: By 1890, 33,000 engineers and chemists were active in the labour force, and by 1910 it was 105,000.[54] Beyond the formal training of the technical staff, however, observers were agreed that the American skilled worker was distinguished by his curiosity, his quick-ness on the uptake and his flexibility.

American mental agility and mobility were helped, no doubt, by the large influx of immigrants coming from various social and historical backgrounds. It was perhaps not entirely accidental, and certainly fortunate that the immigrants included skilled workers from Great Britain and North-West Europe, good farmers from Germany and Scandinavia to fill up the empty spaces to 1890, and potentially unskilled urban labourers from Ireland in the nineteenth century, and from Southern and Eastern Europe after 1890, exactly when they were needed. With immigration exceeding one million in some years, and some 20% of the working force in 1900–1914 foreign born (US Bureau 1960) their impact was considerable. The "Verdoorn" effect, growth in markets permitting growth in productivity, would be one favourable consequence; however, the need for heavy investment, also in roads and housing, to equip these immigrants will have retarded economic growth.

Towards the end of the century, the rise of the "Robber Barons" and of trustification, the power of big business to expand at the expense of consumers, to twist legislatures and judges in its favour has been seen as a factor driving the American economy forward, but at a cost. Today the issue is controversial. Much was wasted, and there was much human cost in exploitation and wide economic fluctuations. At the same time it is difficult to see how else so many people could have been absorbed and so much land opened up so smoothly, while keeping up the fastest rate of per capita growth known to the capitalist West (Hacker 1970).

On the world scene, the USA appeared as a young giant, out-stripping the older industrialized economies relatively and absolutely from about 1880 onward. According to one calculation, her productivity per man-hour, equal to the British in 1890, exceeded hers by almost 25% in 1913; according to another, her real GNP per head, at $1,350 (of 1860) in 1913, was greatly in excess of the British of $1,070, other countries being well below this. The American share of the world's industrial production has been estimated at 23% in 1870, rising to 30% in 1900 and 32–36% in 1913.[55]

In trade, the role of the USA was less prominent, since much of her effort was devoted to developing within her own borders, and in any case, in the early years it was her primary exports which predominated. The share of imports plus exports as a proportion of national income was actually declining with the falling significance of raw cotten exports, from 17% early in the 19th century to 14% by its end (Kuznets

1966). This was a lower proportion than in other countries, though it was still growing faster in absolute terms: the USA accounting for 7% of world trade in 1840, 10% in 1880 and 14% in 1913 (Maddison 1982, Rostow 1978, p. 71). In 1851-60, raw materials still formed 62% of exports and food another 6.6%, as against only 12.3% for finished manufactures; by 1906-10 this had been much reduced, the rates being 32%, 9% and 27% respectively. Similarly, the imports of manufactures fell from 51% to 25% of total imports between these years (US Bureau 1960).

Until 1873, the American merchandise trade balance was negative, it then varied for 20 years and after 1894 became increasingly positive. In the early stages of industrailization, the USA had been an international borrower. Canals and railways from the 1820's, often via state loans, and mining and cattle companies later on, were the chief attractions for foreign, mostly British, investors. The significance of this foreign capital is difficult to gauge. Net foreign investments are estimated at $75 million at the beginning of the century and at $377 million by 1860, which was 3-4% of total capital, but a much higher proportion of joint-stock capital. After the Civil War, foreign capital of quite different proportions came in, net indebtedness reaching almost $2 billion by 1876. Further large inflows were then counterbalanced in part by American investments abroad, mainly in Canada and Latin America. By 1914, some $7 billion of foreign capital was matched by about half that sum. $3.5 million, of US capital abroad (Williamson 1964).

The Role of the State
An understanding of the role of the State in the American industrial-ization process is made difficult by the federal structure of the country. Lobbies of special interests can more easily get their way in the smaller orbit of the individual states, and use the political power of the latter to influence the federal government, in turn. On the whole, the public authorities were more alive to economic development, and more eager to support it, than the more mixed societies of Europe.

This is seen most clearly in the years of the Civil War and the period following when, the southern states being temporarily excluded from power, northern entrepreneurial interests prevailed unhindered. Among the legislation which they inspired, beside the Homestead Act of 1862 already mentioned, and the Legal Tender Act of 1862 and

National Bank Acts of 1863 and 1864 made necessary by the war, there were also the Contract Labour Law of 1864 (which largely remained a dead letter), the decision to subsidise the Pacific Railway, and the raising of the tariff. Taxes were high only in war time. By 1885, the Federal Government contented itself with revenues drawn from tariffs and excise, having abolished income and corporation taxes again as soon as conditions permitted. The support for railways and canals has already been noted.

After the demise of the Second US Bank in 1836 there was no central bank, and between 1832 and 1863 there was also no federal regulation of banking. The banks licenced by the states formed a wide variety of institutions, characterised by their large number and lack of security. It is difficult to believe that the resulting free-for-all could have helped the country's development, though no doubt it made some people very rich.

From the 1860's the federal government attempted to foster a national (i.e. federally licensed) banking system with a legal minimum backing for the note issue; the state banks (i.e. those registered with the states) were taxed and discriminated against in other ways. The latter nevertheless survived in large numbers, though each trade crisis still saw large numbers of bankruptcies, and the banking world remained fragmented and unstable until the outbreak of World War I. There were 1931 banks in existence in 1866 and 27,864 banks in 1914 (US Bureau 1860).

The war-time note issue ("greenbacks") became subject to much lobbying after the Resumption Act of 1875 (resumption 1879), while the silver-producing states, backed by farmer-dominated states which were inclined to favour inflationary policies, wanted to widen the market for the metal by using the widening stream of silver as monetary backing. The resulting compromise added to the metallic base but did not make US prices diverge from world prices. Money supply per head rose from $18.40 in 1860 to $83.20 in 1900.

The multiplicity of banks ensured an adequate supply of short-term finance, mainly by bills, and in some cases also some long-term finance for state governments and railways as well as farm mortgages. For the needs for long-term capital by the larger trusts and companies by the end of the century they were inadequate, and large private investment banks, sometimes acting as syndicates, expanded to fulfil that function. Including foreign capital, there seemed never to be any shortage of capital for the industrialization process in America, but, like much else, it seems to have been provided with a good deal of waste.

Possibly it was the tariff issue in which the American Government showed its pro-industrial bias most clearly. Industrial protection had been the accepted policy while the country was mainly agrarian, but it was continued, at variable rates, even after the infant industries had become dominant giants. In 1861–4 average tariff rates were increased from 25% to 47%. The McKinley Tariff of 1890, raising average rates to 49½%, and the Dingley Tariff of 1897, raising them higher still to 57% (Jones 1965) were passed well after the phase of industrialization. The much less powerful interests in favour of free trade, such as those of cotton producers, merchants and, towards the end of the century, the farmers, could at best modify, but not reverse, the protectionist policy. However, in terms of growth and development, tariffs were probably of minor importance (North 1966).

Finally, the Interstate Commerce Act of 1887 and the Sherman Anti-Trust Act of 1890 were evidence of other clashes of interests. Both were passed, by the pressure of voters, against some of the most active, expansionary but monopolistic industrial interests; but they were rendered largely nugatory by the decisions of the Courts.

3. THE LATECOMERS, AND PROSPECTS FOR THE TWENTIETH CENTURY

3.1. Russia

Russia fits best into the Gerschenkron model of the industrialization of backward economies (Gerschenkron 1963, 1966b), and this is not entirely surprising, since all the indications are that it was developed with that country mainly in mind. To start the catching-up process the very backward economy would, according to Gerschenkron, have to have an exceptionally rapid "spurt", especially in manufacturing, it would quickly acquire large plants and an emphasis on capital goods, but there would be little modernization in agriculture and little improvement in the standard of living, and the process would need much help from the Government. All this did indeed apply to Russia in her first spurt of the 1890's; in the second spurt, 1907–14, much of this was modified, and the banks were substituted for the Government as suppliers of capital and as driving force towards expansion and towards amalgamation, for by then Russia was no longer so extremely "backward" (Gerschenkron 1963, also Rostow 1978, p. 427). The Gerschenkron thesis has received much criticism even when applied to Russia (Gregory 1977). For one thing, it is said, the role of the Government has been exaggerated. The railways, it is true, were fostered by the state, which built the first main line, and owned some three quarters of all lines by 1913 (Kahan 1978). They, in turn, helped to develop an iron and steel capacity, especially in the 1890's, when over half the iron and steel output was supplied to them, as well as engineering skills. By their forward linkages, the railways brought many areas of agrarian as well as mining and industrial production into their needed contact with outside markets, and, expanding from 1,600 km in 1860 to 70,000 km by the outbreak of World War I, thus forming the second largest nerwork in the world, they were a significant element in the Russian growth process.

But there was a cost to this initiative. Much of the taxation with which, according to Gerschenkron, the Government raised the revenue in order to be able to build railways, to provide orders for the armaments industry as well as carry out its other main policy, the stabilization of the currency, was levied on the peasant. The peasantry was squeezed also in order to force it to disgorge its grain for export, the so-called "hunger exports", and this undermined the chance of developing a home market, one of the necessary conditions for successful industrialization. The thesis of the increasingly exploited and impoverished peasantry after the Emancipation of 1861 is also found in much of the standard literature (e.g. Grossman 1973), but has recently been seriously questioned (Simms 1977, Gregory 1982). Many statistics point to a growing productivity in agriculture (see below) rather than to declining consumption: thus output per head of coarse grains rose from an average of 2.21 quarters in 1864-6 to 2.81 in 1900-5.[56]

Another element in Government policy was the tariff. It reached its high point at 33% of the value of imports with the Mendeleev tariff of 1891 before declining again from 1900. It raised the costs for Polish industrialists and for the agrarian sector, which paid most of the taxes, so that the benefit of protectionism turned out to be dubious at best (Bonwetsch 1975). Much effort was devoted by the Government also to set the scene for the introduction of the gold standard, achieved in 1897. The necessary deflation cannot have helped the modernization process, and Russia was one of the few countries to undergo its first industrialization spurt in a deflationary framework (Crisp 1967).

The creation of a modern banking system almost out of nothing was also helped, both by direct Government action and by its sanctioning of private joint-stock banks. The State Bank had the task of keeping the gold reserve and had a note issue monopoly after 1897. It aided the other banks by credits on bills of 6–12 months. By 1914 some dozen joint-stock banks dominated trade and industry and frequently took the initiative in expansion plans, but there were numerous other banks, including many run by local authorities, as well as savings banks (Crisp 1967, Portal 1966). Joint-stock capital rose from 42 million roubles in 1861 to 3,426 million roubles in 1914, or some 80-fold, and the State as well as the banks encouraged the formation of cartels and syndicates.

For a poor country, the savings ratio of Tsarist Russia was astonishingly high, Net investment, as a proportion of net national product, averaged 8.1% in 1885-7, 9.8% in 1899–1901 and 11.3% in 1911. This was far above any other comparable country and approached the rates for Germany and the USA.[57] Nevertheless, there was still a capital shortage, especially for the modern sectors, and one object of stabilizing the currency in 1897 had been to attract foreign investment. In this the Government certainly succeeded. Up to 1900, much of the foreign investment had gone direct to firms, but following the depression of 1900–1906 in which much money was lost, it tended to go to banks in Russia which then channelled it further. Estimates of the amounts involved vary, but it appears that foreign capital as a proportion of joint-stock capital rose from 25% in 1890 to 38% in 1914; in the later years, 1911-13, its share of newly issued capital was as high as 50%. In some sectors the share was much higher still: thus in 1914 it was 90% in mining, 40% in metallurgy, 50% in chemicals, 74% in railways and 43% in banks.[58] A high rate of net capital inflow was achieved in 1892-1901, averaging 150 million roubles, and this was exceeded in 1906-1913, the peak at 578 million roubles being reached in 1913.[59]

A comparison with the USA would show some similarities in the vast extent of the land mass, much of it still unoccupied, in the country's rich mineral resources, in its orientation as a primary product export economy, and in being able to attract foreign capital in a crucial period of growth. But the differences were possibly more significant: the climate was harsher and the soil less fertile (Baykov 1954), transport was more difficult to construct so that distances were more costly to bridge, and the population was very much poorer. Above all, Russia suffered from her feudal past, such as a peasantry still largely tied to the land, a weak entrepreneurial tradition and a heavy-handed bureaucracy dominated by conservative agrarian estate owners. Moreover, she had to appear as a Great Power and waste much of her resources on wars and other trappings which that status required. At the same time, her very size gave Russia a significant place in the world economy despite her poverty. In 1913 she had 8.7% of the world's population, 7.4% of the world's income and 7.6% of the world's industrial potential.[60] She had the largest area under a sugar crop in Europe, for example, and was the world's fourth largest cotton producer.

As one would expect of a backward economy, food processing and textiles formed the most important industrial sectors. Thus among manufacturing and mining, textiles contributed 31.5% of value added in 1887 and 19.2% still in 1913, and food processing 24.4% and 20.9% respectively. But there had been strong developments in iron (Krivoi Rog) and coal (Donetz basin) in the Ukraine, oil in Baku and engineering in several large cities: metallurgy and mining represented 29.5% of value added in 1887 and 51.6% in 1913. Calculated on a different basis, capital goods production rose from 30% in 1890 to 38% in 1913 of all industrial production.[61]

Concentrated in certain very limited key areas (Portal 1966, Milward and Saul 1977), Russian industry bore very strongly the character of a dual economy: huge works of a modern kind existed side by side and were in part supplied by primitive domestic workshops; modern methods, introduced by foreign specialists, requiring frequently a high capital intensity to by-pass the shortage of skilled labour, could be found next to ramshackle and wholly obsolescent plant. Of the two million workers in plants of over 20 employees, no fewer than 40% were in plants employing more than 1000.[62]

It was in agriculture that the Russian economy had its weakest link. In spite of a good deal of progress, it remained in absolute terms far behind European standards in yields and efficiency. Capital equipment was low and primitive, methods antiquated, specialisation extremely limited, and there was an acute shortage of land for the peasants or, putting it differently, a crippling overpopulation in the villages. Yet the peasant had to provide the surplus both for exports and for Government revenue.

According to Gerschenkron (1966a) the Emancipation Edict of 1861, by tying the peasant to his village community, and continuing periodic re-distributions of village lands among its numbers, inhibited the application of innovation in the villages and at the same time prevented peasants from moving to the towns. That the majority of factory workers maintained contacts with their villages, had little skill, and were not fully integrated in their factories is confirmed from other sources. There was also a high degree of illiteracy by European standards, though there was much improvement in this regard in the half-century to 1914 (Crisp 1978). Yet the handicap may have been exaggerated. Cities did grow, and a core of workers was accumulated. At the time of the Emancipation, 4 million peasants had been landless, 600,000 had worked in Government mills and mines and 500,000 in private mills. The number of wage workers increased from 4 million in 1860 to 17½ million in 1913.[63]

As far as the villages were concerned, peasants improved their techniques, and capital used in agriculture rose by 23% between 1891 and 1912. The yields of all grains in the 50 provinces of European Russia rose from an index of 75 in 1861-5 (1888-90 = 100) to 131 in 1911-14.[64] The pressure on the land was eased in part by peasants hiring increasingly quantities of the noble's land, and in part by migration to virgin lands in Siberia, four million having moved from the 1890's to 1914, recalling some of the American experience. The Stolypin reforms, beginning 1906, came too late to make much difference before 1913, but some two million families, many of them among the more enterprising in the village, took the opportunity to claim their share from the village community and start up independently. 2.8 million families, or about one-fifth of the total, had already been outside the communities before that.

Industrial growth rates compared well with those of other industrializing countries. In real terms, industrial output grew at a rate of 5-5½% a year between 1860 and 1913, equivalent to perhaps 3½% a year per head. The rate of growth was very uneven, however, and it is the spurts, with growth rates at 7.6% in 1890-5, 9.2% in 1895-1900 and 7.5% in 1910-13[65] which gave Gerschenkron and others the conviction that an industrial "take-off" was occurring or was about to occur. In certain key sectors the growth rate was much higher. Thus the output of pig iron increased 14-fold between 1860 and 1913, that of coal 77-fold, and the cotton consumption ninefold, while the population rose only 2.36 times.

Agricultural growth rates were much lower than that. While industrial output rose by 124% between 1890 and 1913, grain output rose by 35% only. Nevertheless, the output of the main grains increased 3-fold between 1861 and 1913, of potatoes 7-fold, and of cash crops 6½-fold, altogether just over threefold and therefore ahead of the population increase. In view of its size, the growth of the agricultural sector dominated the overall growth rate. Total commodity production expanded at a rate of the order of 2¼% over the whole period.[66]

The per capita improvement was much less than that because of the exceptionally high rate of population growth of around 1.5% a year. Nevertheless, real GNP per head expanded in this period by 0.9% a year, compared with the average of 1.4% of the leading 15 countries according to Maddison; according to Gregory, the per capita product expanded at the rate of 1.6% a year, a rate faster than any except Japan, while the growth rate of the product per worker, c. 1.1% p.a., was lower than that of most others.[67]

This growth had, however, started from a very low level. Even by 1913, the Russian GNP per head was only about one-third of the British, one-quarter of the American and one-half of the German; in terms of industrial potential the gap was wider still (Table XIV). Thus pig iron production in 1913 per head was one-tenth of the British and one-fourteenth of the American, crude steel production one-seventh and one-twelfth, and raw cotton consumption one-eleventh and one-sixth, respectively. Measuring differently, Rostow found that in 1913

TABLE XIV
GNP and Industrialization per Head, Russia and Japan, 1800–1913[a]

	GNP per head in $ US of 1960			Level of industrialization per head, Index: UK 1900 = 100			
	1830	1860	1913	1800	1860	1880	1913
Russia	180	200	345	6	8	10	20
Japan	180	175	310	7	7	9	20
UK	370	600	1070	16	64	87	115
USA	240	550	1350	9	21	38	126

[a] Sources: Bairoch (1981) p. 10; Bairoch (1982) p. 281.

Russia was 48 years behind the USA in pig iron output, 66 years in coal output and 43 years in cotton fabrics, but was equal in cigarettes.[68] Russian backwardness was shown also in such measurable factors as infantile mortality and illiteracy (Gattrell 1986). Did all this amount to a successful industrialization, a "take-off", before World War I? Rostow himself dated the Russian "take-off" to 1890–1905, the "drive to maturity" then being dated 1905–56. Gregory, following the same line, writes of 1885–1913 as the "industrialization era", and Kahan terms the years from the late 1880's to 1914 the phase of "intensive industrialization". Gerschenkron was inclined to doubt that description, and Goldsmith considered Russia even in 1913 an "undeveloped country". Perhaps Gregory's summary of the views of western scholars comes closest to an acceptable compromise. According to him, "the Russian economy was at best on the threshold of modernization on the eve of the First World War, though it had established many of the preconditions for sustained development".[69]

Alternatively, and perhaps more aptly, the concentrations of modern industry, and even the railway network, may be considered as enclaves, as foreign bodies almost, implanted from outside on an economy which was not yet ready for them and over large areas remained hardly affected by them. Russia had most obviously a dual economy, which in some respects was far behind European standards, but in others, though relatively circumscribed sectors, was close to the industrial and scientific leaders (Pollard 1981, Milward and Saul 1977). Little more than two million true factory workers were swamped by a vast rural population which contributed, even when fishing and forestry is included, little more than half the national income, (Crisp 1978, Falkus 1968, Gregory 1982). The peasantry reduced the averages, as well as the chances of a real breakthrough.

Given her size, it was not perhaps surprising that Russia's exports formed only around 5–6% of GNP. Her links with the world economy consisted largely in exporting foodstuffs and importing not only finished manufactures, but also pig iron and other vital semi-manufactures. Russia was a foreign borrower to a major extent, but it seemed unlikely that in the absence of a war she could have, like the Scandinavian countries, used the foreign capital to solve short-term transitional problems as a start to the rapid expansion of a modern economy. Rather, she looked like the model for the Balkans, Portugal and for most non-European countries: there, active, apparently strong governments and a few impressive urban and industrial centres similarly block the view towards a largely inert, poor, backward agrarian population which would have many years to wait, and many sacrifices to make, before genuine industrialization on a broad basis could take root.

3.2. Japan

Parallels between Russia and Japan readily suggest themselves. Both were deeply traditional societies, shocked out of their stagnation by events in the 1850's and 1860's: Russia by defeat in the Crimean War, followed by the Emancipation Edict of 1861, Japan by the arrival of Commodore Perry, followed by the Meiji restoration in 1868. In both, a rapid growth towards a modern industrial structure set in the mid-1880's. But the contrasts are equally great. Russia covers a vast area and is rich in natural resources, whereas the Japanese people are crowded into narrow strips of land between the mountains and the sea and possess few minerals. Later on, Japan was to try to imitate the western path as far as possible whereas Russia chose deliberately a totally different path of economic planning and the elimination of the private entrepreneur. Above all, Russia was the most backward of the major European states, Japan the most advanced among the non-European communities.

It is Japan's unparalleled economic success in the most recent years which has focussed attention on her early stages of industrialization. A large literature, much of it in western languages, has asked whether there was something in the methods of her start, or even in the pre-existing society, that ensured success far above that of all other similar economies.

There is some controversy regarding the preceding "traditional" Tokugawa era (MacPherson 1987). The population had been almost stationary for over a century, and consisted, according to some, mostly of "unfree, poverty-stricken peasants".[70] Others saw her as a "vigorous and sophisticated, if still traditional and feudal society", or even as a "vigorous, advanced and effective traditional society. In many ways ways it was more advanced than many countries in Africa or Latin America today."[71] In particular, the steady growth of agricultural output and the rise of a rural domestic industry have been

emphasised (Smith, 1959 and 1973). Among the favourable pre-
conditions was a stable political and legal environment, based on a
system of checks and balances; a good infrastructure including a good
road system; a high level of literacy and a tradition of respect for
learning; and an active merchant class not without political influence
and familiar with a sophisticated credit system. There were skilled
craftsmen in the town gilds and in the village proto-industries, and
there had been a slow evolution to wage labour even before 1868 (Taira
1978). The debate has thus turned, to some extent, on whether in the
extremely short period of transition, one could speak of a sharp break
or rather a smooth transition (Inkster 1979).

One way of resolving that issue would be to point out that Japan,
with all its developed social structure, was extremely poor in comparison with western countries at the onset of her industrialization drive.
While western European countries could count, at an equivalent period
in their development, on a per capita income of $220–300 US of 1965
and the USA even on $474 (1834–43), the Japanese level at that time
was put by Kuznets at only $74.[72]

It is unclear how far this was made up by a different structure of
expenditure, such as cheap, if effective wooden houses, and better
social overheads such as a good water supply, efficient soil removal,
and well maintained roads and bridges (Hanley 1983).

As in all traditional societies, agriculture was much the most important sector. No exact data exist for the mid-century, but even in 1887
agriculture supplied 42.5% of the domestic product and 35.5% still in
1911, while employing 73% and 66% respectively of the gainfully
occupied population in 1887 and 1902 (Ohkawa and Shinohara 1979,
Ohkawa and Rosovsky 1973). For 1879/83 Kuznets estimated the
labour force in agriculture at 81.6% of the total (Kuznets 1968) and it
must have been higher still in the 1850's.

Agricultural land was extremely scarce in Japan, amounting to 0.9
hectare per male farm worker in 1880 compared with 8.8 ha in
Denmark, 14.3 ha in Great Britain and 25.4 ha in the USA.[73] Even in
1908, it was still only 1.5 ha, and much of the increase had consisted
in taking in difficult hilly and marginal land (MacPherson 1987).
Increases in output and productivity were therefore difficult to
achieve. Nevertheless, output grew by 1.42% a year in 1887–97, by 1.62
a year in 1897–1904 and by 1.67% a year in 1904–11. Total output rose
by 93% between 1880 and 1915, labour productivity by 97% and land

productivity by 57%, for rice alone by 53%. This was a faster rise than in any country except Canada.[74] It was achieved by better methods, such as double cropping of rice, artificial incubation of silk worms, more animals, better seeds and more manure, by low-cost loans and better incentives after the proportional tax had been changed in 1873 to a fixed tax (Dore 1960).

At the same time, agriculture was able to give up much labour to industry, so that the Japanese industrialization took place with an easy labour supply, despite the relatively slow growth of total population. Net migration from the land ran at half a million in the decade 1875–85 rising to three million in 1910–20, forming 70% of the addition to non-agrarian employment in the 1870's and 83% in the 1910's (MacPherson 1987). This was encouraged by the huge difference in wages, reflecting differences in productivity of between 2:1 and 3:1, between agriculture and the rest (Kelley and Williamson 1974). As a result, there was a substantial productivy gain from the sectoral shift, estimated at 0.55% a year in 1887–97, 0.70% in 1897–1904, 0.01% in 1904–11 and 0.92% in 1911–19.[75]

Higher growth rates, though from a very small start, were achieved by industry, including manufacturing, mining and construction. Between 1887 and the First World War, manufacturing output grew by almost 6% a year (productivity by well over 4%), to raise that sector's contribution to national income from 13.6% to 20.3% in that period. The "facilitating" sector (transport, communications, public utilities), grew by well over 9% a year, and construction by almost 5% (Ohkawa and Shinohara 1979). These are well up to the best of the American and Russian growth rates.

Among the most successful industries was the cotton industry which turned in those years from an importing industry to a net exporter, first of yarn and ultimately also of fabrics, chiefly to China and other Asiatic countries. By 1913 there were 2½ million spindles. Other textiles, including the silk filature also did well, and much of the expansion of this period was carried by the traditional handicraft industries. The heavy industries fared less well, being hampered both by a poor resource base and problems of technological transfer. Shipbuilding did best, and the launching of two large warships of 10,000 tons in 1906 ranked Japan among the leading nations in this field. Against this, iron and steel output did not exceed a quarter million tons by the outbreak of World War I and coal output was a mere 22 million tons. The first railway was not built until 1872, with British capital and foreign engineers. After 1880, when the total was 100 miles, the mileage doubled every three years, and by 1914, some 7,000 miles were in operation (Tsuru 1963). With 1874 = 100, the index of output by value for 1913 at constant prices was: food products 318, textiles 1800, iron and steel 3119 and machinery 7145. The annual rate of growth of horsepower in manufacturing was 10.9% in the 1890's and 17.3% in the 1900's.[76]

The number of industrial workers was growing at a rate of 5-8% a year and had risen to 1.45 million by 1914, of whom 949,000 were classed as factory workers. In the textile mills, the majority of workers were female and badly paid. Elsewhere, too, the labour surplus and low opportunity wage on the farm made recruitment and disciplining easy. Some labour unrest occurred in the early years of the twentieth century, but it was in those years the management learnt to develop the family spirit, the lifetime employment and the seniority system which later were to become a hallmark, and a possible cause of success, of large-scale Japanese industry (Nakagawa 1979, Hirschmeier 1968, Taira 1978). Strong elements of a "dual economy" were visible in Japan as in other countries at that stage.

The data for the tertiary sector are the least reliable, but the overall growth rate may be taken as reasonably accurate. According to the latest estimates, GDP rose by 3.21% a year in 1887-97, by 1.83% a year in 1897-1904 and by 3.30% a year in 1904-19; per capita the growth rates in the same period were 2.25%, 0.67% and 2.11% respectively (Ohkawa and Shinohara 1979). Maddison's estimates, extending over 1879-1913, but worked out on a basis allowing international comparisons, were slightly lower: 2.7% a year for GNP and 1.7% for the per capita figure, which were below the USA rates but above those of the other countries listed (Maddison 1969, p. 31). Whether the period 1885-1905 was the "take-off", followed by the drive to technical maturity, which Rostow professed to see, will depend on one's view of the low absolute level; Ohkawa and Rosovsky called c. 1885-1900 the phase of "initial" modern economic growth.[77]

Nothing is known about the rate of saving and capital formation in the early Meiji years. From 1887 onward, when estimates exist, net savings as proportion of Net National Product remained stable at c. 6%, while gross savings showed a rising tendency from 13% in 1887 to c. 16% of GNP in 1908-12. Investment remained stable at c. 11% of

GNP, with additional military investments rising from 2% to 6% in the same years.[78]

These rates would correspond to Rostow's "taking-off" phase, though Japan is pictured as an economy with abundant labour but a shortage of capital. When, after the turn of the century, more had to be invested in modern plant as well as in the expanding Government sector, Japan had recourse to foreign loans. Before 1900 these had been kept to a minimum and quickly repaid, but following the two successful wars against China and Russia as well as the establishment of the gold standard in 1897, loans on relatively easy terms became possible, mainly to the Japanese Government which then channeled them to private sectors. By 1913, the outstanding foreign debt of Japan had reached c. £200 million.

The growth rate of Japanese exports, at c. 8% a year, was much the fastest among all major economies. In quantity, exports rose from an index of 1.4 (1960 = 100) in 1873-7 to 26.1 in 1913-17, or about 18-fold, while imports rose similarly from an index of 1.6 to 31.2.[79] In the early Meiji years, the emerging adverse balance was covered by the export of specie; at the end of our period, some foreign borrowing helped the balance. Among exports, primary products fell from 42.5% by value in 1874-83 to 12.3% in 1907-16, while textiles rose from 42.4% to 53.6%. Among imports, manufacturers fell from 91.2% to 50.0%, but within this total, heavy manufactures, especially machines, rose from 19.4% to 34.4% in the same years.[80] Japan did not regain her tariff autonomy before 1911, and protection therefore played a minor role in her pre-war development.

The role of the Japanese Government in these development has been much disputed. It is generally agreed that it provided a stable political and legal environment after feudal obligations had been abolished, the Samurai bought off and intertrade freed following the Meiji restoration. It not only prevented the colonisation of Japan, but actually turned Japan into a colonial power. The indemnity which it exacted from China in 1895-8 became the basis of the gold standard and may thus be considered an aid, but the total effect of Japan's militarism was probably a negative one for Japan's economy (Kelley and Williamson 1974).

Of great significance was the heavy taxation of the peasants from whom in the early years as much as 80% of the State's revenue was raised, falling to 25% by 1900. With this income the Government improved the infrastructure, financed the first railways, guaranteed the interest on the major shipping line, supported armanents industries including shipyards and ironworks, fostered the first pilot cotton and other textile mills and helped the rise of that peculiar Japanese financially/trading/industrial conglomerate, the *zaibatsu*. It is a moot point how far the bureaucracy, many members of which came from the demoted *samurai* class, helped the spirit of enterprise among Japan's industrial innovators along with their xenophobic nationalism.

The early recognition by the Government of the importance of education is undisputed. The decree on general education was issued in 1872. By 1896, 96% of the population between 6 and 13 was registered in a primary school. Modern University teaching began in 1870 and the Imperial University of Tokyo was opened in 1877. The Government was also instrumental in sending Japanese students and technicians abroad, and bringing foreign specialists to Japan.

The early Meiji Governments tended to be inflationary, but after the Matsukata deflation of 1881-5, more stable conditions prevailed. The National Banking Acts 1872 and 1876 were modelled on the USA, and an adequate banking system developed. By 1880 there were 153 national banks. There were also "quasi-banks", lending short-term for production and cash crops, and private banks. The *zaibatsu* generally provided their own finance. Joint-stock organisation was made easy by the law of 1890 (Patrick 1967, Yamamura 1978).

The industrialization of Japan is sometimes seen as having occurred in its first forty years or so, much of the growth as well as the export balance were produced by the primary sector. Keeping the income of the primary sector down allowed the innovating sectors to grow successfully. But the process was circumscribed by the limited growth potential of the primary sector. In Japan's case it was fortunate that when nothing more could be squeezed from it, the industrial and commercial sectors were large enough to carry economic growth forward largely on their own (Ohkawa and Rosovsky 1973, Lockwood 1968). An alternative model would see growth until 1885 carried by all the traditional sectors, from which the modern ones took over thereafter.

Following the Rostow model, exports have been termed the leading sector, and it has indeed been held that "foreign trade has been indispensable to Japan's industrial growth".[81] The problem here is that up to the 1890's, exports accounted for only 11% of the increase in output, and even thereafter this rose only to 25% (Klein and Ohkawa 1968, p. 95). Against this, it has been held that Japan's industrial growth was due to supply rather than demand, with import substitution especially of cotton goods, and substitution of manufacturing for primary production playing the key role (Klein and Ohkawa 1968, p. 112).

Apart from the speed of development, the main difference between Japan and the European and North American industrializing countries is to be found in Japan's ability to carry through similar stages of structural change and technical innovation at a much lower absolute level of per capita income. In part this may be due to technical factors, such as wooden instead of brick and stone houses, and rice instead of wheat and oats. It may also have to do with an earlier alternative social development which laid greater stress on social discipline, on community spirit and on the development of the self by study and adaptation than was the case among the Christian societies in the West. Whatever the cause, it has helped Japan to secure export markets by paying lower wages than other countries using similar technical equipment.

3.3. Into the Twentieth Century

There were other countries, beside those described here which laid the foundations for industrialization before the nineteenth century was over. These included the Netherlands, the Scandinavian countries, the British Dominions and certain circumscribed regions in Austria-Hungary, in Italy and possibly also in Spain. Each of these would show some special features of their own which might diverge from the models presented here. All of them developed into fully industrialized and modernised countries or regions in the early decades of the twentieth century.

Interestingly enough there were then for a considerable time no newcomers to the industrial scene. With the doubtful exceptions of enclave economies in Manchuria, China, India, and possibly Argentina and Brazil, the countries in which industrialization had not taken root before 1900, showed no development in that direction in the following decades either, until the end of the Second World War. It was as if the world industrialization process had run out of steam, or alternatively, as if the countries that remained backward after 1900 lacked too many of the preconditions to make the jump over the ever higher threshold.

When a new industrialization wave began after 1945, it was to be on a different basis, and with a logic different from that of the earlier models. The state took a much more decisive role, attempting in each case to force the whole economy into a mould that was considered favourable to industrialization. An enormous range of technologies had meanwhile become available, and these were often made freely available by the richer advanced countries. Steps were taken consciously, on the advice of numerous specialists in the economic, social and technical fields, which in the nineteenth century had not had the benefit of any, or at mostly only a minimal, State direction. It did not follow that the industrialization drives of the later twentieth century were any more successful than those of the nineteenth: rather the contrary. But the result has been that the lessons of the various models of the nineteenth century have become of very limited value for the progress of today's industrializers.

NOTES

1. Cameron (1985) p. 9.
2. From the Introduction, cited in Cameron (1985) p. 1.
3. Morris and Adelman (1988) use 35 classificatory indicators for 23 countries in three periods each.
4. An interesting variant alleges that British (and American) agricultural productivity was higher than that of the European continent, not so much because of better technology, but because the Anglo-Saxons worked harder (Clark, 1987).
5. "All countries which are now relatively developed have at some time in the past gone through a period of rapid acceleration in the course of which their rate of annual net investment has moved from 5 per cent or less to 12 per cent or more. This is what we mean by an Industrial Revolution." (Lewis, 1955, p. 208).
6. Table IIA above.
7. Samuel (1977) p. 18.
8. Contributions of the main export items to the total growth in exports, at current prices, have been calculated to have been as follows (in per cent of total exports):

	1784–6 to 1814–6	1814–6 to 1844–6
woollen yarns and cloth	14.0	10.8
cotton yarns and cloth	52.8	45.9
other textiles	6.9	9.8
iron and iron goods		16.9
other metals, raw and wrought	7.9	16.6
other goods	18.3	(Crouzet, 1985, p. 180).

9. Crouzet (1972c) p. 98.
10. "The English industrial revolution took place before ours. It is useful not to forget this. The great majority of the technical means that we applied had already been invented and put to work — England having borne the costs of the first tryouts. It would thus be absurd to think of the Belgian industrial revolution — and of all industrial revolutions other than the English — as if they were like the English industrial revolution." (Lebrun, 1979, pp. 64-5 [Transl. S.P.]).
11. Crouzet (1972c) p. 98.
12. Caron (1979) p. 35.
13. Cameron (1985) p. 13.
14. O'Brien and Keyder (1978) p. 179.
15. Price (1981) p. 49.
16. O'Brien and Keyder (1978) pp. 92, 94. A number of other statistical sources give other absolute figures, but the general tendency is the same. Thus the share of those employed in agriculture in the total employed population has been put as 51.9% in 1840-5 and 49.8% still in 1866. (Caron, 1979, p. 33).
17. Price (1983) pp. 304, 346. Also see Milward and Saul (1973) p. 355.
18. Cameron (1967) p. 115.
19. Cameron (1967) p. 128.
20. Caron (1979) p. 97.
21. Braudel and Labrousse (1976) pp. 307, 314.
22. Palmade (1972) p. 175; Levy-Leboyer (1978) p. 239; Cameron (1961) Tables II and III, pp. 85, 88.
23. Clapham (1963) p. 96.
24. The performance of the major sectors is shown in Table IX.
25. Kuznets (1971) pp. 38-9.
26. Clapham (1963) p. 279.
27. Hoffmann (1965) pp. 22, 44, 143, 147.
28. Hoffmann (1965) pp. 59, 63.
29. Henning (1984) p. 218.
30. Annual incomes per head in 1913, in Marks:

Hamburg	1313
Berlin	1254
Brandenburg	962
Westphalia	735
Baden	710
East Prussia	486
West Prussia	480
Posen	465

(Borchardt, 1973, p. 139).
31. Landes (1969) p. 330.
32. Fremdling (1977) p. 584.
33. 89% of the monetary circulation was metallic in 1850, and 56% still in 1913. (Hoffmann, 1965, pp. 814-5).
34. Hoffmann (1965) p. 773.
35. Borchardt (1973) p. 138.
36. Hoffmann (1965) p. 262.
37. Borchardt (1973) pp. 102-3.
38. Bairoch (1965) pp. 1096, 1102, 1107, 1108; Bairoch (1982) pp. 275, 281, 294, 296.
39. Rosenberg (1972) p. 85.
40. Rostow (1978) p. 384; Temin (1975) p. 64.
41. North (1966) pp. 680, 693; Hacker (1970) p. 69; Jones (1965) p. 145.

42. Rosenberg (1972) p. 60.
43. Investment in canals in $ million:
    Annual Average 1820–5   2.2
    Annual Average 1826–37   4.8
    Annual Average 1838   12.3
    Annual Average 1839   13.6
    Annual Average 1840   14.3
    Annual Average 1841   11.7
    Annual Average 1842–50   2.9
    (Williamson, 1964, p. 279).
44. Rostow (1978) pp. 151–2.
45. Fogel (1964) pp. 132, 196.
46. Rostow (1978) p. 137.
47. In 1860, the South had 35% of the population, but only 13% of the manufacturing of the United States.
48. Gallman (1960) p. 31; Lebergott (1984) p. 301; Parker (1971) p. 177.
49. U.S. Bureau (1960) p. 284; Towne and Rasmussen, pp. 282, 292–3; also Kendrick (1961) p. 365.
50. Rostow (1984) p. 164.
51. Herfindahl (1966) pp. 294, 297.
52. Gallman (1960) p. 83; also Kuznets (1966) pp. 136–7; Rosenberg (1972) pp. 44–6.
53. Maddison (1982) pp. 44–5.
54. Kendrick (1961) p. 90; also Lebergott (1984) p. 349; Hacker (1970) pp. 212ff.
55. Maddison (1982) pp. 98, 112; Bairoch (1981) p. 10 and (1982) pp. 275, 296; North (1966) p. 673.
56. Pollard (1981) p. 197.
57. Gregory (1982) pp. 56–7, 172.
58. Portal (1966) p. 851; Gattrell (1986) p. 208; Bovykin (1975) p. 194; Crisp (1967) p. 227; Nötzold (1975) p. 240.
59. Gregory (1982) pp. 97–8.
60. Bairoch (1982) p. 296.
61. Bovykin (1975) p. 203; Goldsmith (1961) p. 459; Falkus (1968) p. 58.
62. Milward and Saul (1977) p. 416. Also Gattrell (1986) pp. 10, 154–5.
63. Maddison (1969) p. 90; Gattrell (1986) p. 85.
64. Gattrell (1986) p. 101; Kahan (1978) p. 301. Also Simms (1977).
65. Crisp (1967) p. 184; Nötzold (1975) p. 235; Goldsmith (1961) pp. 442, 465; Grossman (1973) p. 489; Rostow (1978) p. 428.
66. Goldsmith (1961) pp. 446, 471; Kahan (1978) p. 269.
67. Maddison (1969) p. 31; Gregory (1982) p. 162; Goldsmith (1961) p. 474.
68. Gregory (1982) p. 155; Bairoch (1965) pp. 1102, 1104; Rostow (1978) pp. 434–5.
69. Rostow (1978) p. 437; Gregory (1982) pp. 1, 166; Kahan (1978) p. 265; Gerschenkron (1963) pp. 163–4; Goldsmith (1961) p. 443; Gregory (1977) p. 215.
70. Lockwood (1968) p. 3.
71. Rostow (1978) p. 416; Ohkawa and Rosovsky (1973) p. 7.
72. Kuznets (1971) p. 24, also Kuznets (1968) p. 200. But see Table XIV.
73. Yamada and Hayami (1979) pp. 94–5.
74. Yamada and Hayami (1979) p. 91; Ohkawa and Shinohara (1979) p. 38; Klein and Ohkawa (1968) p. 157; Maddison (1969) p. 89.
75. Ohkawa and Shinohara (1979) p. 44
76. Ohkawa and Shinohara (1979) pp. 105, 302–3; Minami (1977) p. 939.
77. Rostow (1978) p. 423; Ohkawa and Rosovsky (1973) p. 9.
78. Ohkawa and Shinohara (1979) pp. 18, 30.

79. Kunio (1979) p. 55; Maddison (1969) p. 29.
80. Yamazawa and Yamamoto (1979) p. 135.
81. Yamazawa and Yamamoto (1979) p. 134.

REFERENCES

Abramovitz, Moses and David, Paul A. (1973), "Reinterpreting Economic Growth: Parables and Reality," *American Economic Review, Papers and Proceedings*, 63, 428-39.

Allen, G.C., (1966), "The Industrialization of the Far East," pp. 875-923 in Habakkuk and Postan.

Ashton, Thomas S. (1948), *The Industrial Revolution 1760-1830*. London: Oxford University Press.

Ashton, Thomas S. (1959), *Economic Fluctuations in England 1700-1800*. Oxford: Clarendon.

Ashworth, William (1977), "Typologies and Evidence: Has Nineteenth-Century Europe a Guide to Economic Growth?" *Economic History Review*, 2nd Ser. 30, 140-158.

Bairoch, Paul (1965), "Niveaux de développement économique de 1810 à 1910," *Annales E.S.C.*, 20, 1091-1117.

Bairoch, Paul (1976), "Europe's National Product 1800-1975," *Journal of European Economic History*, 5, 273-340.

Bairoch, Paul (1981), "The Main Trends in National Economic Disparities since the Industrial Revolution," pp. 3-17 in Bairoch and Lévy-Leboyer.

Bairoch, Paul (1982), "International Industrialization Levels from 1750 to 1980," *Journal of European Economic History*, 11, 269-310.

Bairoch, Paul and Lévy-Leboyer, Maurice, eds (1981), *Disparities in Economic Development Since the Industrial Revolution*. London: McMillan.

Barkin, Kenneth D (1970), *The Controversy over German Industrialization 1890-1902*. Chicago: University Press.

Baykov, Alexander (1954), "The Economic Development of Russia," *Economic History Review*, 7, 137-49.

Beck, Thomas D. and Beck, Martha W. (1987), *French Notables: Reflections of Industrialization and Regionalism*. New York: Lang.

Berend, Ivan T and Ranki, György (1982), *The European Periphery and Industrialization 1780-1914*. Cambridge: University Press.

Bergier, Jean-François (1983), *Die Wirtschaftsgeschichte der Schweiz*. Zurich: Benziger.

Bloch, Marc (1953), "Toward a Comparative History of European Societies," pp. 494-521 in Lane, Frederic C. and Riemersma, Jelle C., eds., *Enterprise and Secular Change*. London: Allen & Unwin.

Bonwetsch, Berndt, (1975), "Handelspolitik und Industrialisierung. Zur aussenwirtschaftlichen Abhängigkeit Russlands," pp. 277-99 in Geyer.

Bovykin, Valerij I., (1975), "Probleme der industriellen Entwicklung Russlands," pp. 188-209 in Geyer.

Borchardt, Knut, (1968), "Regionale Wachstumsdifferenzierung in Deutschland im 19. Jahrhundert," pp. 115-130 in Lütge, Friedrich, ed., *Wirtschaftliche und soziale Probleme der gewerblichen Entwicklung im 15.-16. und 19. Jahrhundert*. Stuttgart: Fischer.

Borchardt, Knut, (1973), "Germany 1700-1914," pp. 76-160 in Cipolla.

Braudel, Fernand and Labrousse, Ernest, eds., (1976), *Histoire économique et sociale de la France. Tome III: L'avènement de l'ère industrielle (1789 - années 1880).* Paris: Presses universitaires.

Brown, E.H Phelps, and Browne, Margaret H., (1968), *A Century of Pay.* London: MacMillan.

Cameron, Rondo E., (1958), "Economic Growth and Stagnation in France, 1815-1914," *Journal of Modern History* **30**: 1-13.

Cameron, Rondo, (1961), *France and the Economic Development of Europe.* Princeton: University press.

Cameron, Rondo, ed., (1967), *Banking in the Early Stages of Industrialization.* New York and London: Oxford University Press.

Cameron, Rondo, ed., (1970), *Essays in French Economic History.* Homewood: Irwin.

Cameron, Rondo, ed., (1972), *Banking and Economic Development: Some Lessons of History.* New York and London: Oxford University Press.

Cameron, Rondo, (1985), "A New View of European Industrialization," *Economic History Review,* 2nd Ser. **38**: 1-23

Cameron, Rondo and Freedeman, Charles E., (1983), "French Economic Growth: A Radical Revision," *Social Science History* **7**: 3-30.

Caron, François, (1970), "Railway Investment 1850-1914,", pp. 315-340 in Cameron.

Caron, François, (1979), *An Economic History of Modern France.* London: Methuen.

Chambers, J. D., (1953), "Enclosure and Labour Supply in the Industrial Revolution", *Economic History Review,* 2nd Ser. **5**: 319-43.

Chandler, Alfred D., Jr., (1959), "The Beginning of 'Big Business' in American Industry," *Business History Review* **33**: 1-31.

Chandler, Alfred D., Jr., (1977), *The Visible Hand. The Managerial Revolution in American Business.* Cambridge, Mass.: Belknap.

Chenery, Hollis B., (1960), "Patterns of Industrial Growth," *American Economic Review* **50**: 624-654.

Chenery, Hollis B. and Syrquin, Moises, (1975), *Patterns of Development, 1950-1970.* London: Oxford University Press.

Cipolla, Carlo, ed (1973), *Fontana Economic History of Europe,* vol. 4. *The Emergence of Industrial Societies.* London: Collins-Fontana.

Clapham, John H., (1963), *The Economic Development of France and Germany 1815-1914.* Cambridge: University Press. (First edition 1921).

Clark, George N., (1953), *The Idea of the Industrial Revolution.* Glasgow: University Press.

Clark, Gregory, (1987), "Productivity Growth without Technical Change in European Agriculture before 1850," *Journal of Economic History* **47**: 419-432.

Cochran, Thomas C., (1981), *Frontiers of Changes: Early Industrialism in America.* New York: Oxford University Press.

Coleman, Donald C., (1956), "Industrial Growth and Industrial Revolutions," *Economica* N.S. **23**: 1-22.

Coelho, Philip R.P., (1973), "The Profitability of Imperialism: The British Experience in the West Indies 1768-1772", *Explorations in Economic History* **10**: 253-280.

Conference on Research in Income and Wealth, (1960), *Trends in the American Economy in the Nineteenth Century.* Studies in Income and Wealth, vol. 24. Princeton: University Press.

Conference on Research in Income and Wealth, (1966), *Output, Employment and Productivity in the United States After 1800.* Studies in Income and Wealth, vol. 30. New York: Columbia University Press.

Conrad, Alfred H. and Meyer, John R., (1971), "The Economics of Slavery in the Ante-Bellum South," pp. 342-361 in Fogel and Engerman.

Craeybeckx, Jan, (1970), "The Beginnings of the Industrial Revolution in Belgium," pp. 187-200 in Cameron.

Crafts, N.F.R., (1984), "Economic Growth in France and Britain, 1830-1910. A Review of the Evidence," *Journal of Economic History* **44**: 49-67.

Crafts, N.F.R. (1985a), "Industrial Revolution in England and France: Some Thoughts on the Question 'Why Was England First?'" pp. 119-136 in Mokyr 1985a.

Crafts, N.F.R., (1985b), "Income Elasticities of Demand and the Release of Labor by Agriculture During the British Industrial Revolution. A Further Appraisal", pp. 151-163 in Mokyr 1985a.

Crafts. N.F.R., (1985c), *British Economic Growth During the Industrial Revolution*. Oxford: Clarendon.

Crafts N.F.R., (1987), "British Economic Growth, 1700-1850: Some Difficulties of Interpretation," *Explorations in Economic History* **24**: 245-268.

Crisp, Olga, (1967), "Russia, 1860-1914," pp. 183-238 in Cameron.

Crisp, Olga, (1978), "Labour and Industrialization in Russia," pp. 308-415 in Mathias and Postan.

Crouzet, François, (1967), "England and France in the Eighteenth Century. A Comparative Analysis of two Economic Growths", pp. 139-174 in Hartwell, R.M., *The Causes of the Industrial Revolution in England*. London: Methuen.

Crouzet, François, (1970), "An Annual Index of French Industrial Production in the 19th Century", pp. 245-278 in Cameron.

Crouzet, François, ed. (1972a), *Capital Formation in the Industrial Revolution*. London: Methuen.

Crouzet, François, (1972b), "Capital Formation in Great Britain During the Industrial Revolution", pp. 162-222 in Crouzet, 1972a.

Crouzet, François, (1972c), "Western Europe and Great Britain: 'Catching up'in the First Half of the Nineteenth Century," pp. 98-125 in Youngson, A.J., *Economic Development in the Long Run*. London: Allen & Unwin.

Crouzet, François, (1985), *De la superiorité de l'Angleterre sur la France: L'Economique et l'Imaginaire XVIIe-XXe siècles*. Paris: Perrin.

David, Paul, (1971), "The Mechanization of Reaping in the Ante-Bellum Midwest,", pp. 214-227 in Fogel and Engerman.

Deane, Phyllis, (1965), *The First Industrial Revolution*. Cambridge: University Press.

Deane, Phyllis, (1972), "Capital Formation in Britain Before the Railway Age", pp. 94-118 in Crouzet 1972a.

Deane, Phyllis and Cole, W.A., (1967), *British Economic Growth 1688-1959*. Cambridge: University Press.

Dhont, Jan and Bruwier, Marinette, (1973), "The Low Countires 1700-1914", pp. 329-366, in Cipolla.

Dore, R.P., "Agricultural Improvement in Japan (1870-1900)," *Economic Development and Cultural Change* **9**: 69-91.

Easterlin, Richard, (1960), "Interregional Differences in Per Capita Income, Population and Total Income, 1840-1950," pp. 73-140 in Conference on Income and Wealth.

Economic History, First International Conference, (1960), *Contributions*. Paris and Hague: Mouton.

Eicher, C. and Witt, L., (1964), *Agriculture in Economic Development*. New York: McGraw-Hill.

Eistert, Ekkehart, (1970), *Die Beeinflussung des Wirtschaftswachstums in Deutschland von 1883 bis 1913 durch das Bankensystem*. Berlin: Duncker & Humblot.

Eliasberg, Vera F., (1966), "Some Aspects of Development in the Coal Mining Industry, 1839-1918," pp. 405-435 in Conference on Income and Wealth.

Engerman, Stanley L, (1972), "The Slave Trade and British Capital Formation in the Eighteenth Century: A Comment on the Williams Thesis", *Business History Review* **46**: 430-443.

Eversley, David E.C., (1967), "The Home Market and Economic Growth in England 1750-80," pp. 206-59 in Jones, Eric L. and Mingay, G.E., eds., *Land, Labour and Population in the Industrial Revolution*. London: Arnold.

Falkus, M.E., (1968), "Russia's National Income, 1913: a Revaluation," *Economica* N.S. **35**: 52-73.

Fei, John C.H. and Ranis, Gustav, (1969), "Economic Development in Historical Perspective,", *American Economic Review* **59**: 386-400.

Feinstein, Charles H., (1978), "Capital Formation in Great Britain," pp. 28-96 in Mathias and Postan.

Feinstein, Charles H. and Pollard, Sidney, eds., (1988), *Studies in Capital Formation in the United Kingdom 1750-1920*. Oxford: Clarendon.

Feis, Herbert, (1964), *Europe, the World's Banker 1870-1914*. New York: Kelley.

Findlay, R., (1982), "Trade and Growth in the Industrial Revolution", pp. 178-188 in Kindleberger, Charles P. and di Tella, Guido, eds., *Economics in the Long View: Essays in Honour of W.W. Rostow*, vol. 1. New York: New York University Press.

Fischer, Wolfram, (1963), "Government Activity and Industrialization in Germany," pp. 83-94 in Rostow.

Fishlow, Albert, (1971), "Levels of Nineteenth-Century American Investment in Education", pp. 265-73 in Fogel and Engerman.

Floud, Roderick and McCloskey, Donald N., eds., (1981), *The Economic History of Britain Since 1700, vol. 1, 1700-1860*. Cambridge: University Press.

Fogel, Robert William, (1970), *Railroads and American Economic Growth*. Baltimore: Johns Hopkins.

Fogel, Robert William and Engerman, Stanley L., eds., (1971), *The Reinterpretation of American Economic History*. New York: Harper & Row.

Fogel, Robert William and Engerman, Stanley L., (1974), *Time on the Cross: The Economics of American Negro Slavery*. 2 vols. Boston: Little, Brown.

Fohlen, Claude, (1970), "The Industrial Revolution in France", pp. 201-225 in Cameron.

Fohlen, Claude, (1971), *Qu'est-ce que la révolution industrielle?* Paris: Laffont.

Fohlen, Claude, (1973), "The Industrial Revolution in France 1700-1914," pp. 7-75, in Cipolla.

Fohlen, Claude, (1978), "Entrepreneurship and Management in France in the Nineteenth Century," pp. 347-381 in Mathias and Postan.

Forberger, Rudolf, (1982), *Die industrielle Revolution in Sachsen 1800-1861*. Berlin: Akademie-Verlag.

Fores, Micheal, (1981), "The Myth of a British Industrial Revolution," *History* **66**: 181-198.

Fremdling, Rainer, (1977), "Railroads and German Economic Growth: A Leading Sector Analysis with a Comparison to the United States and Great Britain," *Journal of Economic History* **37**: 583-604

Fremdling, Rainer, (1985), *Eisenbahnen und deutsches Wirtschaftswachstum 1840-1879*. Dortmund: Gesellschaft f. westfälische Wirtschaftsgeschichte.

Fremdling, Rainer and O'Brien, Patrick K., eds., (1983), *Productivity in the Economies of Europe*. Stuttgart: Klett-Cotta.

Fremdling, Rainer and Tilly, Richard, (1976), "German Banks, German Growth and Economic History," *Journal of Economic History* **36**: 416-424.

Fremdling, Rainer and Tilly, Richard H., eds., (1979), *Industrialisierung und Raum*.

Studien zur regionalen Differenzierung im Deutschland des 19. Jahrhunderts. Stuttgart: Klett-Cotta.

Gallman, Robert E., (1960), "Commodity Output, 1839-1899," pp. 13-67 in Conference on Income and Wealth.

Gallman, Robert E., (1966), "Gross National Product in the United States, 1834-1909," pp. 3-76 in Conference on Income and Wealth.

Gattrell, Peter, (1986), *The Tsarist Economy 1850-1917.* New York: St. Martin's Press.

Geary, Frank, (1988), "Balanced and Unbalanced Growth in XIXth Century Europe," *Journal of European Economic History* 17: 349-357.

Gerschenkron, Alexander, (1963), "The Early Phases of Industrialization in Russia", pp. 151-69 in Rostow 1963a.

Gerschenkron, Alexander, (1966a), "Agrarian Policies and Industrialization: Russia 1861-1917", pp. 706-800 in Habakkuk and Postan.

Gerschenkron, Alexander, (1966b), *Economic Backwardness in Historical Perspective.* Cambridge, Mass.: Harvard University Press.

Gever, Dietrich, ed., (1975), *Wirtschaft und Gesellschaft im vorrevolutionären Russland.* Cologne: Kiepenheuer & Witsch.

Gilboy, Elizabeth W., (1932), "Demand as a Factor in the Industrial Revolution," pp. 620-639 in Cole, A.H. et al, eds., *Facts and Factors in Economic History: Articles by Former Students of Edwin Francis Gay.* Cambridge, Mass.: Harvard University Press.

Goldin, Claudia D. and Frank D. Lewis, (1975), "The Economic Costs of the American Civil War: Estimates and Implications," *Journal of Economic History* 35: 299-326.

Goldsmith, Raymond W., (1961), "The Economic Growth of Tasrist Russia 1860-1913," *Economic Development and Cultural Change* 9: 441-75.

Goldsmith, Raymond W., (1969), *Financial Structure and Economic Development.* New Haven: Yale University Press.

Gould, J.D., (1972), *Eonomic Growth in History: Survey and Analysis.* London: Methuen.

Gregory, Paul R., (1977), "Russian Industrialization and Economic Growth," *Jahrbücher für Geschichte Osteuropas* N.F. 25: 200-18.

Gregory, Paul R., (1982), *Russian National Income, 1885-1913.* Cambridge: University Press.

Grossman, Gregory, (1973), "Russia and the Soviet Union", pp. 486-531 in Cipolla.

Habakkuk, H.J., (1962), *American and British Technology in the Nineteenth Century.* Cambridge: University Press.

Habakkuk, H.J. and Postan, M., eds., (1966), *Cambridge Economic History of Europe,* vol. VI. Cambridge: University Press.

Haberler, G. and Stern, R.M., (1962), *Equilibrium and Growth in the World Economy: Economic Essays by Ragnar Nurkse.* Cambridge, Mass.: Harvard University Press.

Hacker, Louis M., (1968), *The World of Andrew Carnegie: 1865-1901.* Philadelphia: Lippincott.

Hacker, Louis M., (1970), *The Course of American Economic Growth and Development.* New York: Wiley.

Hanley, Susan, (1983), "A High Standard of Living in Nineteenth-Century Japan: Fact or Fantasy?" *Journal of Economic History* 43: 183-92.

Harley, C. Knick, (1982), "British Industrialization Before 1841: Evidence of Slower Growth During the Industrial Revolution," *Journal of Economic History* 42: 267-289

Hartwell, R.M., (1965), "The Causes of the Industrial Revolution: An Essay in Methodology," *Economic History Review,* 2nd Ser. 18: 164-82.

Heim, Carol E. and Mirowski, Philip, (1987), "Interest Rates and Crowding-Out during

Britain's Industrial Revolution," *Journal of Economic History* **47**: 117–139.

Henderson, William O., (1954), *Britain and Industrial Europe*. Liverpool: University Press.

Henderson, William O., (1967), *The Industrial Revolution on the Continent: Germany, France, Russia, 1800–1914*. London: Cass.

Henderson, William Otto, (1975), *The Rise of German Industrial Power 1834–1914*. London: Temple Smith.

Henning, Friedrich–Wilhelm, (1984), *Die Industrialisierung in Deutschland 1800 bis 1914*. Paderborn: Schöningh. (first edition: 1973)

Herfindahl, Orris C., (1966), "Development of the Major Metal Mining Industries in the United States from 1839 to 1909," pp. 293–346 in Conference.

Heywood, Colin, (1981), "The Role of the Peasantry in French Industrialization, 1815–1880," *Economic History Review*, 2nd Ser. **34**: 359–376.

Hirschman, Albert O., (1958), *The Strategy of Economic Development*. New Haven: Yale University Press.

Hirschmeier, Johannes, (1968), *The Origins of Entrepreneurship in Meiji Japan*. Cambridge, Mass.: Harvard University Press.

Hoffmann, Walther G., (1958), *The Growth of Industrial Economies*. Manchester: University Press.

Hoffmann, Walther G., (1963), "The Take-off in Germany," pp. 95–118 in Rostow 1963a.

Hoffmann, Walther G., (1965), *Das Wachstum der deutschen Wirtschaft seit der Mitte des 19. Jahrhunderts*. Berlin: Springer.

Hoffmann, Walther D, and Müller, J.H., (1959), *Das deutsche Volkseinkommen 1851–1957*. Tübingen: Mohr.

Hounshell, David A., (1984), *From the American System to Mass Production 1800–1932*. Baltimore: Johns Hopkins.

Hughes, Jonathan R.T., (1970), *Industrialization in Economic History*. New York: McGraw-Hill.

Inikori, Joseph E., (1987), "Slavery and the Development of Industrial Capitalism in England", *Journal of Interdisciplinary History* **17**: 771–793.

Inkster, Ian, (1979), "Meiji Economic Development in Perspective: Revisionist Comments upon the Industrial Revolution in Japan," *The Developing Economies* **17**:

Jones, Eric L., (1981), *The European Miracle. Environments, Economies and Geopolitics in the History of Europe and Asia*. Cambridge: University Press.

Jones, Eric L., (1988), *Growth Recurring: Economic Change in World History*. Oxford: Clarendon.

Jones. E.L. and Woolf, S.J., (1969), "The Historical Role of Agrarian Change in Economic Development," pp. 1–21 in Jones, E.L. and Woolf S.J., *Agrarian Change and Economic Development*. London: Methuen.

Jones, Peter d'A., (1965), *The Consumer Society. A History of American Capitalism*. Harmondsworth: Penguin.

Kaelble, Hartmut, (1983), "Der Mythos von der rapiden Industrialisierung in Deutschland," *Geschichte und Gesellschaft* **9**: 106–18.

Kahan, Arcadius, (1978), "Capital Formation in the Period of Early Industrialization in Russia, 1890–1913," pp. 265–307 in Mathias and Postan.

Kelley, Allen C. and Williamson, Jeffrey G., (1974), *Lessons from Japanese Development*. Chicago: University Press.

Kemp, Tom, (1971), *Economic Forces in French History*. London: Dobson.

Kendrick, John W., (1961), Productivity Trends in the United States. Princeton: University Press.

Klein, Lawrence and Ohkawa, Kazushi, (1968), *Economic Growth: the Japanese Experience since the Meiji Era*. Homewood: Irwin.

Kunio, Yoshihara, (1979), *Japanese Economic Development. A Short Introduction*. Oxford: University Press.

Kuznets, Simon, (1951), "The State as a Unit in the Study of Economic Growth," *Journal of Economic History* 11: 25-41.

Kuznets, Simon, (1966), *Modern Economic Growth*. New Haven: Yale University Press.

Kuznets, Simon, (1968), "Trends in Level and Structure of Consumption," pp. 197-242 in Klein and Ohkawa.

Kuznets, Simon, (1971a), "Notes on the Pattern of U.S. Economic Growth," pp. 17-24 in Fogel and Engerman.

Kuznets, Simon, (1971b). *The Economic Growth of Nations. Total Output and Production Structure*. Cambridge Mass.: Belknap.

Landes, David S., (1959), "French Entrepreneurship and Industrial Growth in the Nineteenth Century", *Journal of Economic History* 9: 45-61.

Landes, David S., (1969), *The Unbound Prometheus: Technological Change and Industrial Development in Western Europe from 1750 to the Present*, Cambridge: University Press.

Lebergott, Stanley, (1966), "Labor Force and Employment 1800-1960", pp. 117-204 in Conference on Income and Wealth.

Lebergott, Stanley, (1984), *The Americans: An Economic Record*. New York: Norton.

Lebovics, Herman, (1967), "'Agrarians' versus 'Industrializers': Social Conservative Resistance to Industrialism and Capitalism in Late Nineteenth Century Germany," *International Review of Social History* 12: 31-65.

Lebrun, Pierre, (1979), et al. *Histoire quantitative et développement de la Belgique*, II.1, *Essai sur la révolution industrielle en Belgique, 1770-1847*. Liège.

Leet, Don R. and Shaw, John A., (1978), "French Eonomic Stagnation, 1700-1960: Old Economic History Revisited," *Journal of Interdisciplinary History* 8: 531-544.

Léon, Pierre, (1960), "L'industrialisation en France en tant que facteur de croissance économique du debut du XVIIIe siècle à nos jours," pp. 163-204, in Economic History.

Léon, Pierre, (1974), "Structure du commerce extérieur et évolution industrielle de la France à la fin du XVIIIe siècle," pp. 407-432 in Braudel, Fernand et al., eds., *Conjoncture économique et structures sociales: Hommage à Ernest Labrousse*. Paris: Mouton

Lévy-Leboyer, Maurice, (1968), "Les processus d'industrialisation: le cas de l'Angleterre et de la France," *Revue Historique* 239: 281-298.

Lévy-Leboyer, Maurice, (1976), "Innovation and Business Strategies in Nineteenth- and Twentieth-Century France,", pp. 87-135 in Carter, Edward C. II, et al. éds., *Enterprise and Entrepreneur in Nineteenth- and Twentieth-Century France*. Baltimore: Johns Hopkins.

Lévy-Leboyer, Maurice, ed., (1977), *La position internationale de la France*. Paris: École des Hautes Études en Sciences Sociales.

Lévy-Leboyer, Maurice, (1978), "Capital Investment and Economic Growth in France, 1820-1930," pp. 231-295 in Mathias and Postan.

Lévy-Leboyer, Maurice and Bourguignon, François, (1985), *L'économie française au XIXe siècle: Analyse macro-économique*. Paris: Economica.

Lewis, W. Arthur, (1954), "Economic Development with Unlimited Supplies of Labour," *Manchester School of Economic and Social Studies* 22: 139-191.

Lewis, W. Arthur, (1955), *The Theory of Economic Growth*. London: Allen & Unwin.

Lewis, W. Arthur, (1958), "Unlimited Labour: Further Notes," *Manchester School of Economic and Social Studies* 26: 1-32.

Lincke, Bruno, (1910), *Die schweizerische Maschinenindustrie und ihre Entwicklung in wirtschaftlicher Beziehung*. Zurich: University Diss.

Lindert, Peter H. and Williamson, Jeffrey G., (1985), "English Workers' Living Standard During the Industrial Revolution: A New Look," pp. 177-205 in Mokyr, 1985a.

Lockwood, William W., (1968), *The Economic Development of Japan. Growth and Structural Change*. Princeton: University Press.

MacPherson, W.J., (1987), *The Economic Development of Japan c. 1868-1941*. London: MacMillan.

Maddison, Angus, (1969), *Economic Growth in Japan and the USSR*. London: Allen & Unwin.

Maddison, Angus, (1982), *Phases of Capitalist Development*. Oxford: University Press.

Mahaim, Ernest, (1905), "Les débuts de l'établissement John Cockerill à Seraing," *Vierteljahrschrift für Social- und Wirtschaftsgeschichte* 3: 627-648.

Mantoux, Paul, (1961), *The Industrial Revolution in the Eighteenth Century: An Outline of the Beginnings of the Modern Factory System in England*. London: Cape. First edition: 1928.

Marczewski, Jean, (1963), "The Take-Off Hypothesis and French Experience", pp. 119-138 in Rostow.

Marczewski, Jean, (1965), "Le produit physique de l'économie française de 1789 à 1913 [comparison avec la Grande-Bretagne]", *Cahiers de l'I.S.E.A.* Ser. AF 4, **163**: VII-CLIV

Markovitch, T.J., (1965), "L'industrie française de 1789 à 1964 - Sources et methodes," *Cahiers de l'I.S.E.A.* Ser AF 4, **163**: 1-231.

Mathias, Peter, (1983), *The First Industrial Nation: An Economic History of Britain 1700-1914*. London: Methuen.

Mathias, Peter and Postan, M.M., eds, (1978). *The Cambridge Economic History of Europe*, vol. VII, *The Industrial Economies: Capital, Labour and Enterprise*, 2 Parts. Cambridge: University Press.

Mendels, Franklin F., (1972), "Proto-Industrialization: The First Phase of the Industrialization Process," *Journal of Economic History* 32: 241-61.

Milward, Alan S. and Saul, B.S., (1973), *The Economic Development of Continental Europe 1780-1870*. London: Allen & Unwin.

Milward, Alan S. and Saul, B.S., (1977), *The Development of the Economies of Continental Europe 1850-1914*. London: Allen & Unwin.

Minami, Ryoshin, (1977), "Mechanical Power in the Industrialization of Japan," *Journal of Economic History* 37: 935-58.

Mitchell, Brian Redman, (1981), *European Historical Statistics: 1750-1975*. London: MacMillan.

Mokyr, Joel, (1976), *Industrialization in the Low Countries 1795-1850*. New Haven: Yale University Press.

Mokyr, Joel, ed., (1985a), *The Economics of the Industrial Revolution*. Totowa: Rowan and Allanheld.

Mokyr, Joel, (1985b), "Demand vs. Supply in the Industrial Revolution," pp. 97-118 in Mokyr 1985a.

Mokyr, Joel, (1987), "Has the Industrial Revolution Been Crowded Out? Some Reflections on Crafts and Williamson," *Explorations in Economic History* **24**: 293-319.

Mokyr, Joel, (1988), "Is there still Life in the Pessimist Case? Consumption During the Industrial Revolution 1790-1850," *Journal of Economic History* **48**: 69-92.

Moore, Wilbert E., (1979), *World Modernization: the Limits of Convergence*. New York: Elsevier.

Morris, Cynthia Taft and Adelman, Irma, (1988), *Comparative Patterns of Economic Development 1850-1914*. Baltimore: Johns Hopkins University Press.

Mottek, Hans, (1964), *Wirtschaftsgeschichte Deutschlands, Ein Grundriss*. Band II. *Von der Zeit der französichen Revolution bis zur Zeit der Bismarckschen Reichsgründung*. Berlin (East): Deutscher Verlag der Wissenschaften.

Musson, A.E., (1978), *The Growth of British Industry*. London: Batsford.

Musson, A.E. and Robinson, Eric, (1969), *Science and Technology in the Industrial Revolution*. Manchester: University Press.

Nakagawa, Keiichiro, ed., (1979), *Labour and Management. Proceedings of the Fourth Fuji Conference*. Tokyo: University Press.

Nardinelli, Charles, (1988), "Productivity in Nineteenth Century France and Britain: a Note on the Comparison," *Journal of European Economic History* 17: 427-434.

Neuburger, Hugh and Stokes, Houston H., (1974), "German Banks and German Growth: An Empirical View," *Journal of Economic History* 34: 710-731.

Newell, William H., (1973), "The Agricultural Revolution in Nineteenth-Century France,", *Journal of Economic History* 33: 697-731.

Nishikawa, Shunsaku, (1978), "Productivity, Subsistence and By-Employment in the Mid-Nineteenth Century Choshu," *Explorations in Economic History* 15: 69-83.

Nötzold, Jürgen, (1975), "Agrarfrage und Industrialisierung am Vorabend des ersten Weltkrieges," pp. 228-51 in Geyer.

North, Douglass C., (1955), "Location Theory and Regional Economic Growth," *Journal of Political Economy* 63: 243-258.

North, Douglass C., (1961), *The Economic Growth of the United States 1790-1860*. Englewood Cliffs: Prentice-Hall.

North, Douglass, (1966), "Industrialization in the United States", pp. 673-705 in Habakkuk and Postan.

Nye, John Vincent, (1987), "Firm Size and Economic Backwardness: A New Look at the French Industrialization Debate," *Journal of Economic Hitory* 47: 649-669.

O'Brien, Patrick K., (1977), "Agriculture and the Industrial Revolution," *Economic History Review* 30: 166-181.

O'Brien, Patrick, ed., (1983), *Railways and the Economic Development of Western Europe, 1830-1914*. London: MacMillan.

O'Brien, Patrick K., (1986), "Do We Have a Typology for the Study of European Industrialization in the XIXth Century?" *Journal of European Economic History* 15: 291-333.

O'Brien, Patrick and Keyder, Caglar, (1978), *Economic Growth in Britain and France 1780-1914*. London: Allen & Unwin.

Ohkawa, Kazushi and Rosovsky, Henry, (1973), *Japanese Economic Growth. Trend Acceleration in the Twentieth Century*. Stanford: University Press.

Ohkawa, Kazushi and Shinohara, Miyohei, eds., (1979), *Patterns of Japanese Economic Development: a Quantitative Appraisal*. New Haven: Yale University Press.

Palmade, Guy P., (1972), *French Capitalism in the Nineteenth Century*. Newton Abbot: David & Charles. (Original French edition: 1961).

Parker, William N., (1971), "Productivity Growth in American Grain Farming: an Analysis of its 19th Century Sources," in Fogel and Engerman.

Patrick, Hugh T., (1967), "Japan 1868-1914", pp. 239-89 in Cameron.

Perloff, Harvey S., Dunn, Edgar S. Jr., Lampard, Eric. E. and Muth, Richard F., (1960), *Resources and Economic Growth*. Baltimore: Johns Hopkins University Press.

Plessis, Alain, (1987), "Le 'retard' français: la faute à la banque?" pp. 199-210 in Fridenson, Patrick and Straus, André, eds. *Le capitalisme français XIXe-XX siècle*. Paris: Fayard.

Pollard, Sidney, (1972), "Fixed Capital in the Industrial Revolution in Britain", pp. 145-161 in Crouzet 1972a.

Pollard, Sidney, (1973), "Industrialisation and the European Economy," *Economic History Review*, 2nd Ser. **26**: 636-648.

Pollard, Sidney, (1978), "Labour in Great Britain," pp. 97-179 in Mathias and Postan.

Pollard, Sidney, ed., (1980), *Region und Industrialisierung*. Göttingen: Vandenhoeck & Ruprecht.

Pollard, Sidney, (1981), *Peaceful Conquest: The Industrialization of Europe 1760-1970*. Oxford: University Press.

Portal, Roger, (1966), "The Industrialization of Russia," pp. 801-872 in Habakkuk and Postan.

Porter, Roy and Teich, Mikulas, eds., (1986), *Revolution in History*. Cambridge; University Press.

Pounds, Norman J.G., (1985), *An Historical Geography of Europe 1800-1914*. Cambridge: University Press.

Price, Roger, (1981), *An Economic History of Modern France, 1730-1914*. London: MacMillan.

Price, Roger, (1983), *The Modernization of Rural France: Communications Networks and Agricultural Market Structures in Ninetheenth-Century France*. London: Hutchinson.

Ranis, Gustav and Fei, John C.H., (1961), "A Theory of Economic Development," *American Economic Review* **51**: 533-565.

Ransom, Roger L., (1964), "Canals and Development: A Discussion of the Issue," *American Economic Review*, Papers and Proceedings **54**: 365-76.

Ransom, Roger L. and Sutch, Richard, (1977), *One Kind of Freedom: The Economic Consequences of Emancipation*. Cambridge: University Press.

Rist, Marcel, (1970), "Free Trade", pp. 286-314, in Cameron.

Roehl, Richard, (1976), "French Industrialization, a Reconsideration," *Explorations in Economic History* **13** 233-281.

Rosenberg, Nathan, ed., (1969), *The American System of Manufactures*. Edinburgh: University Press.

Rosenberg, Nathan, (1972), *Technology and American Growth*. New York: Harper & Row.

Rosenberg, Nathan, (1976), *Perspectives on Technology*. Cambridge: University Press.

Rostow, Walt W., (1960), *The Stages of Economic Growth. A Non-communist Manifesto*. Cambridge: University Press.

Rostow, Walt W., ed., (1963a), *The Economics of Take-off into Sustained Growth*. London: MacMillan.

Rostow, Walt W., (1963b), "Leading Sectors and the Take-off," pp. 1-21, in Rostow 1963a.

Rostow, Walt. W., (1978), *The World Economy: History and Prospect*. Austin and London: University of Texas Press.

Rosovsky, Henry, (1961), *Capital Formation in Japan 1868-1940*. Glencoe: Free Press.

Rosovsky, Henry, (1966), "Japan's Transition to Modern Economic Growth, 1868-1885," pp. 91-139 in Rosovsky, Henry, ed., *Industrialization in Two Systems*. New York: Wiley.

Samuel, Raphael, (1977), "The Workshop of the World: Steam Power and Hand Technology in mid-Victorian Britain," *History Workshop* **3**: 6-72.

Schumpeter, Joseph A., (1934), *The Theory of Economic Development*. Cambridge, Mass.: Harvard University Press.

Sewell, William H. Jr., (1967), "Marc Bloch and the Logic of Comparative History," *History and Theory* **2**: 208-218.

Simms, James Y., Jr., (1977), "The Crisis in Russian Agriculture at the End of the Nineteenth Century: A Different View," *Slavic Review* **36**: 377-98.

"Slavery as an Obstacle to Economic Growth in the United States. A Panel Discussion." (1967), *Journal of Economic History* **27**: 518-60.

Smith, Thomas C., (1959), *The Agrarian Origins of Modern Japan*. Stanford: University Press.

Smith, Thomas C., (1973), "Pre-Modern Economic Growth: Japan and the West," *Past and Present* **60**: 127-60.

Söderberg, Johan, (1985), "Regional Economic Disparity and Dynamics, 1840-1914: A Comparison between France, Great Britain, Prussia and Sweden," *Journal of European Economic History* **14**: 273-296.

Sombart, Werner, (1927-8), *Der Moderne Kapitalismus*, 3 vols.: Munich: Duncker & Humblot.

Spiethoff, Arthur, (1953), "Pure Theory and Economic Gestalt Theory; Ideal Types and Real Types," pp. 444-463 in Lane, Frederic C. and Riemersma, Jelle C., eds., *Enterprise and Secular Change*. London: Allen & Unwin.

Taira, Koji, (1978), "Factory Labour and the Industrial Revolution in Japan," pp. 166-214 in Mathias and Postan.

Temin, Peter, (1964), *Iron and Steel in Nineteenth-Century America. An Economic Enquiry*. Cambridge, Mass: MIT Press.

Temin, Peter, (1966), "Steam and Waterpower in the Early Nineteenth Century," *Journal of Economic History* **26**: 187-205.

Temin, Peter, (1975), *Causal Factors in American Economic Growth in the Nineteenth Century*. London: MacMillan.

Temin, Peter, (1976), "The Post-Bellum Recovery of the South and the Cost of the Civil War," *Journal of Economic History* **36**: 898-907.

Tilly, Richard, (1967), "Germany 1815-1970", pp. 151-82 in Cameron.

Tilly, Richard H., (1978), "Capital Formation in Germany in the Nineteenth Century," pp. 382-441 in Mathias and Postan.

Tipton, Frank B, (1976), *Regional Variations in the Economic Development of Germany During the Nineteenth Century*. Middletown, Conn.: Wesleyan University Press.

Towne, Marvin W. and Rasmussen, Wayne D., (1960), "Farm Gross Product and Gross Investment in the Nineteenth Century", pp. 255-312 in Conference.

Toynbee, Arnold, (1884), *Lectures on the Industrial Revolution in England*. London: Rivington.

Trebilcock, Clive, (1981), *The Industrialization of the Continental Powers 1780-19141*. London: Longman.

Tsuru, Shigeto, (1963), "The Take-off in Japan, 1868-1900", pp. 139-50 in Rostow 1963a.

Tunzelmann, G.N. von, (1985), "The Standard of Living Debate and Optimal Economic Growth", pp. 207-226 in Mokyr 1985a.

Tyszynski, H., (1951), "World Trade in Manufactured Commodities, 1899-1950," *Manchester School* **19**: 272-304.

U.S. Bureau of Census, (1960), *Historical Statistics of the United States. Colonial Times to 1957*. Washington: Government Printing Office.

Vial, Jean, (1967), *L'industrialisation de la sidérurgie française, 1814-1864*. Paris: Mouton.

Wallerstein, Immanuel, (1974), *The Modern World System: 1. Capitalist Agriculture and the Origins of the European World-Economy in the Sixteenth Century*. London: Academic Press.

Wallerstein, Immanuel, (1980), *The Modern World System: 2. Mercantilism and the*

Consolidation of the European World Economy 1600–1750. New York: Academic Press.

Wehler, Hans-Ulrich, (1987), *Deutsche Gesellschafts-Geschichte*, vol. 1, *1700–1815*, vol. 2, *1815–1848/9*. Munich: Beck.

White, H.D., (1933), *The French International Accounts 1880–1914*. Cambridge, Mass.: Harvard University Press.

Williamson, Geoffrey G., (1964), *American Growth and the Balance of Payments 1820–1913*. Chapel Hill: University of North Carolina Press.

Williamson, Jeffrey G., (1965), "Regional Inequality and the Process of National Development: a Description of the Patterns," *Economic Development and Cultural Change* 13: 3–84.

Williamson, Jeffrey G., (1984), "Why Was British Growth so Slow during the Industrial Revolution?" *Journal of Economic History* 44: 687–712.

Williamson, Jeffrey G., (1985), "The Historical Context of the Classical Labour Surplus Model," *Population and Development Review* 2: 171–191.

Williamson, Jeffrey G., (1987), "Debating the British Industrial Revolution," *Explorations in Economic History* 24: 269–292.

Wright, Gavin, (1974), "Cotton Competition and the Post-Bellum Recovery of the American South," *Journal of Economic History* 34: 610–635.

Wrigley, A.E., (1961), *Industrial Growth and Population Change*. Cambridge: University Press.

Wrigley, E.A., (1986), "Men on the Land and Men in the Countryside: Employment in Agriculture in Early Nineteenth-Century England," pp. 295–336 in Bonfield, L., Smith, R.M. and Wrightson, K., eds., *The World We Have Gained: Histories of Population and Social Structure*. Oxford: Blackwell.

Wrigley, E.A., (1988), *Continuity, Chance and Change: The Character of the Industrial Revolution in England*. Cambridge: University Press.

Yamada, Saburo and Hayami, Yujiro, (1979), "Agriculture," pp. 85–121 in Ohkawa and Shinohara.

Yamamura, Kozo, (1978), "Entrepreneurship, Ownership and Management in Japan," pp. 215–64 in Mathias and Postan.

Yamazawa, Ippei and Yamamoto, Yuzo, (1979), "Trade and Balance of Payments", pp. 134–56 in Ohkawa and Shinohara.

Zamagni, Vera, (1978), *Industrializzione e squilibria regionali in Italia*. Bologna: Il Mulino.

Zevin, Robert Brooke, (1971), "The Growth of Cotton Textile Production after 1815," pp. 122–147 in Fogel and Engerman.

Zunkel, Friedrich, (1962), *Der rheinisch-westfälische Unternehmer 1834–79*. Cologne-Opladen: Westdeutscher Verlag.

INDEX

FUNDAMENTALS OF PURE
AND APPLIED ECONOMICS
SECTIONS AND EDITORS

BALANCE OF PAYMENTS AND INTERNATIONAL FINANCE
W. Branson, Princeton University
DISTRIBUTION
A. Atkinson, London School of Economics
ECONOMIC DEVELOPMENT STUDIES
S. Chakravarty, Delhi School of Economics
ECONOMIC HISTORY
P. David, Stanford University, and M. Lévy-Leboyer, Université
Paris X
ECONOMIC SYSTEMS
J.M. Montias, Yale University
ECONOMICS OF HEALTH, EDUCATION, POVERTY AND
CRIME
V. Fuchs, Stanford University
ECONOMICS OF THE HOUSEHOLD AND INDIVIDUAL
BEHAVIOR
J. Muellbauer, University of Oxford
ECONOMICS OF TECHNOLOGICAL CHANGE
F.M. Scherer, Harvard University
EVOLUTION OF ECONOMIC STRUCTURES, LONG-TERM
MODELS, PLANNING POLICY, INTERNATIONAL ECONOMIC
STRUCTURES
W. Michalski, O.E.C.D., Paris
EXPERIMENTAL ECONOMICS
C. Plott, California Institute of Technology
GOVERNMENT OWNERSHIP AND REGULATION OF
ECONOMIC ACTIVITY
E. Bailey, Carnegie-Mellon University, USA
INTERNATIONAL ECONOMIC ISSUES
B. Balassa, The World Bank
INTERNATIONAL TRADE
M. Kemp, University of New South Wales

FUNDAMENTALS OF PURE AND APPLIED ECONOMICS

PUBLISHED TITLES

Further titles in preparation
ISSN: 0191-1708